# Kaman
# H-43

# Kaman

# H-43

# An Illustrated History

## Wayne Mutza

**Schiffer Military/Aviation History**
Atglen, PA

# ACKNOWLEDGMENTS

This book is the product of many individuals, some of whom participated in the events documented here. I am most fortunate that they shared my enthusiasm about a historical account of the H-43 helicopter. Had they not stirred their memories and blown the dust off memorabilia long since relegated to dark storage spaces, I would still be mired in the early stages of research. It is my fervent hope that this book reflects their exemplary efforts.

My heartfelt thanks to Regina Burns, U.S. Army Aviation Museum; Allen S. Conrad; Dick Van Allen; Ernest Neville; John W. Christianson; Ron Garcia SSGT, USAF; Stan Schaetzle; Paul E. McAllister, Air Combat Command History Office; Edward L. White; Charles J. Burton; Jack L. King; John A. Harman; Bill Hayes; James L. Butera; Martin P. Jester; Bob Collom; William A. Lyell; Jack M. Friell; David Hansen; Tom Hansen; Lennart Lundh; Terry Love; Lawrence M. Smail, U.S. Army TRECOM; Nick Williams; Dale Mutza; William Luther; Joe Ballinger; Carlton R. Damonte; Kevin Lederhos, Timber Choppers Inc.; Marco Dini Bruno, and Skip Robinson.

And to Debbie, whose endless support and wisdom fuels the inspiration for my work.

# DEDICATION

This book is respectfully dedicated to the H-43 crewmen who lost their lives during the honorable performance of their duty

Book Design by Ian Robertson.

Copyright © 1998 by Wayne Mutza
Library of Congress Catalog Number: 97-81408

Printed in China.
ISBN: 0-7643-0529-8

We are interested in hearing from authors with book ideas on related topics.

Published by Schiffer Publishing Ltd.
4880 Lower Valley Road
Atglen, PA 19310
Phone: (610) 593-1777
FAX: (610) 593-2002
E-mail: schifferbk@aol.com
Please write for a free catalog.
This book may be purchased from the publisher.
Please include $3.95 postage.
Try your bookstore first.

# CONTENTS

# Introduction

The date was 1 August 1956. The carrier USS TARAWA was at sea launching a squadron of Douglas Skyraiders. An AD-5 was bridled into the catapult as the pilot prepared himself for the takeoff, bracing for the expected jolt and rapid acceleration. It never came. The launch was a "cold shot," Navy slang for a faulty catapult that produced only enough power to literally shove an airplane off the deck. The lumbering Skyraider crashed into the sea and broke apart, its fuel burning furiously on the water. Luckily, the pilot got out.

A speck in the background got larger as it closed the distance to the pilot, bobbing near the floating inferno. Tarawa's crew crowded the deck as they watched the HOK-1 helicopter loom into view, using its rotor downwash to blow away flames and rescue the pilot with its hoist. The pilot was in the water only 50 seconds.

A few years later, an F-105 Thunderchief declared an emergency inbound to Seymour-Johnson Air Force Base in North Carolina. An H-43B "Huskie" helicopter was "scrambled," which followed the fighter as it made its approach. As the F-105 touched down, its fuselage broke in two just behind the cockpit and both halves burst into flames. The chopper landed just long enough to drop off a fire suppression kit and two firefighters, then hovered to push back the flames.

Spraying foam from a hose, a rescueman waded through the flames as he fought his way to the trapped pilot. As he neared the cockpit, he heard the pilot screaming, "Get out of here, you'll never get me out alive!" A few seconds later, the pilot's shoulder straps were slashed and he was freed from his burning aircraft, suffering only superficial burns. Less than two minutes had passed since he impacted the runway.

Some years later, in Southeast Asia during the 1960s, an unarmed and unarmored HH-43B "Pedro" helicopter responded to a desperate plea for help from a beleaguered Marine unit which had traded blows with the enemy. A badly wounded Marine, who would bleed to death if he was not quickly evacuated, became the Pedro crew's objective. Although enemy groundfire slammed into the hovering helicopter, which lost all its engine oil, the tenacious pilot held his position until the Marine was safely aboard.

In a second operation, the same pilot flew his HH-43 repeatedly through heavy automatic weapons fire into a steep jungle box canyon controlled by the enemy. During two days of sorties into the enemy lair, the persistent Pedro crew succeeded in rescuing the crew of an Army helicopter shot down trying to extricate a five-man patrol surrounded by three enemy platoons. Had it not been for the Pedro crew's valiant actions, those trapped on the ground would have met with death or capture.

All of these are actual events which epitomize the helicopter, its crew and its mission. Often overlooked is the fact that the H-43 has its roots in the Navy, where the time-honored "plane guard" role was pioneered. Later, an improved turbine-powered version joined the Air Force Air Rescue Service, whose global achievements meant the difference between life and death for thousands.

At a time when designers and engineers wrestled with the technical problems of creating helicopters, Kaman Aircraft developed an ingenious design. The awkward configuration sacrificed glamour for efficiency, but helicopters never get much respect anyway. Nor are they sexy. They don't take off with an afterburning roar and disappear in seconds. And they vibrate and whip up cyclonic gusts. Ascribing to those characteristics, and more, the Huskie may not evoke images of grace and splendor; but to those who accepted the challenge of rotary-wing flight every time they climbed aboard the peculiar machine, or to those who knew that the approaching helicopter's distinctive sound meant help, the Huskie was every bit a star performer, and a thing of beauty.

I saw my first Huskie at Bien Hoa Air Base, South Vietnam, in 1971, during a short stint there as an advisor. With most of my time in Bell Hueys, I was naturally curious about the odd-looking helicopter. After shop talk with the crew (which was tricky since the detachment's canine mascot closely guarded the pad), I was left with admiration for the chopper—and a deeper curiosity which I vowed to someday satisfy. Researching the fascinating aircraft for this long overdue account has rendered that impression of 1971 indelible. Read on and discover why.

# Chapter 1
# Birth of a Concept

"Flying Mixmaster, Synchropter, Huskie, Pedro" and even "A sparrow beating itself to death" were terms used to describe the Kaman 600 series helicopter, which ultimately became best known as the H-43. Though an ugly duckling among the world's classic aircraft, the H-43 has more than earned its place in the annals of aviation history. Its development centers around the determination of aeronautical engineer Charles H. Kaman, who forged his diverse corporation in 1945, equipped only with meager funds, a few loyal employees and great persistence. An in-depth treatment of the H-43 would be incomplete without first examining early Navy and Marine Corps use, and the stellar achievements of Kaman, the man and the company.

Though serious attempts at vertical flight date back to the 1800s, it wasn't until 1939 that creative genius Igor Sikorsky met the complexities of rotary-wing flight head-on with the flight of his VS-300. As Sikorsky was hard at work during World War II developing his fling-wing marvel, Kaman stepped from the shadows to prove his theories of helicopter design. At that time, helicopters were unstable, prompting

Kaman to concentrate on stability and control forces. He was convinced that aerodynamic servo flaps on rotor blades would stabilize the rotor and make the craft easier to fly. He also envisioned the use of intermeshing rotors, which promised a huge power saving and greater efficiency over tail rotor designs. Sikorsky was totally committed to the tail rotor configuration, which countered main rotor torque, but tapped precious power without providing lift.

Kaman's "synchropter" configuration used very long rotors which produced large-diameter rotor discs. From the very start, Kaman equipped his helicopters with solid spruce rotor blades, which had natural resilience to obtain torsional deflection. The practice of all-wood blades continued until 26 January 1961, when the first-built H-43B made the first flight with composite rotor blades.

Firm in his belief that a servo flap (also called a servo tab) was the answer to stability problems, Kaman designed small elevator-like flaps, which were attached to the trailing edge of each rotor blade, three-fourths of the way out to the blade tip. Similar to ailerons—the movable surface on the

**Advertising leaflet from Kaman's early period.**

**As a cropduster, the K-190 proved ideal for downwashing insecticide onto plants. (Kaman)**

The first production HTK-1 for the Navy is flanked by a similar commercial machine. A large ventral/dorsal fin was later added to the center tail section. (NMNA)

This was the only Kaman helicopter purchased by the U.S. Coast Guard. Kaman's designation system for its helicopters corresponded to the horsepower rating of the aircraft's typical powerplant. The K-225's bare framework left little room for the required USCG Markings. (U.S. Coast Guard)

trailing edge of an airplane's wing—servo flaps were linked to the pitch controls and changed pitch by utilizing aerodynamic forces acting on the blade itself. Up to that time, helicopter blade pitch was controlled by using mechanical force at the rotor hub to move the entire blade. Being much smaller than the blade, the servo flap required much less force to move and thereby control pitch. That translated to less vibration and easier pilot control, which increased pilot efficiency.

When designing his first helicopter, Kaman focused on German aerodynamicist Anton Flettner's development of the FL-282 helicopter, which used synchronized intermeshing rotors. The contra-rotating rotors were mounted atop pylons which were canted slightly to each side. Not only did the configuration nullify torque, its low rotor disc loading made it very efficient at high altitudes and provided tremendous lifting ability. Blade stall and settling with power, problems that plagued single-rotor helicopters, were nonexistent in Kaman's efficient design. During autorotation—controlling and landing the aircraft with the engine off—the machine settled to earth with the ease of a falling maple seed.

Recognizing the potential for rotary-wing application to military, agricultural and industrial service, Kaman quickly became a prudent businessman and promoter. To satisfy initial investors, Kaman Aircraft met its first challenge by getting

The HTK-1 Bureau Number 128657 was extensively modified to become the world's first twin-turbine helicopter. Here, the aircraft makes a sharp flare on landing during 1956. (Boeing via NMNA)

The electrically-operated rescue hoist could be installed on either side of the HOK-1. The copilot's seat could be positioned rearward for operating the hoist, which could lift 600 pounds. (U.S. Army)

The number of troops the HOK-1 could transport was limited to four - three in the back and one occupying the copilot's seat. (U.S. Army)

The HOK-1 Bureau Number 129835 was evaluated by the Army at Fort Rucker, Alabama during late 1956. Here, it lifts a Howitzer gun carriage. (Kaman)

the first helicopter, the K-125, into the air just before a 15 January 1947 deadline. Faced with an uncertain future, yet confident in their technology and engineering, the budding company turned to building cropdusters. A modified variant, labeled the K-190, first flew in April 1948 and was granted a CAA certificate during April 1949 for commercial use. Shortly thereafter, an improved version with a 225 horsepower engine, called the K-225, was also certified. During 1949, K-225s conducted insect and plant disease control flights along the entire Atlantic seaboard, demonstrating the effectiveness of the twin-rotor configuration in aerial spray work.

Over the long run, however, cropdusting with helicopters proved costly and dangerous. Kaman knew they had to stir military interest if they expected to stay in business. The company gained a foothold in the aviation industry by selling the U.S. Navy two Model K-225s, while a single unit was delivered to the Coast Guard. The machines were powered by a Lycoming 0-435-A2 piston engine, which propelled them to a top speed of 72 mph. During March 1950 a prototype was demonstrated to a group of skeptical officials at the Naval Air Test Center. A total of eleven K-225s were built.

During that period, the U.S. Marine Corps expressed their interest in acquiring an observation helicopter that could

First flight of the HOK-1 which was experimentally fitted with a Lycoming gas turbine engine during 1956. (Kaman)

A H-43B from the first production batch was put on display at the Pentagon during March 1959. (Kaman)

Weapons mounted to the H-43B were offered to the Air Force as an option during the late 1950s. Barely visible at the pilot's station of this early H-43B is a crude sight for the quad .30 cal. machine guns. (Kaman)

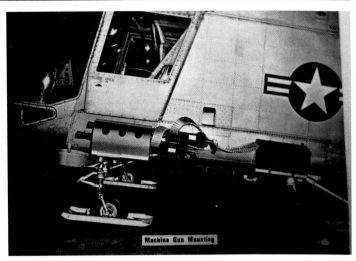

This .30 cal. machine gun arrangement was one of a few made available to the Air Force with the H-43B package. (Kaman)

double as a rescue craft and utility vehicle. Though the larger firms of Sikorsky and Bell submitted proposals for the design, Kaman won the competition as a result of the Navy's successful evaluation of the K-225. That led to the companies' first big production contract in June 1950 for the HOK-1, which stood for "Helicopter, Observation, Kaman—Model One." The aircraft could carry four persons or two litter patients, plus attendant and pilot in a fully enclosed airframe.

In an effort to narrow the large time gap between delivery of the original K-225s and the first HOK-1 delivery, and shore up its finances in the process, Kaman convinced the Navy to purchase trainers based on the K-225. The contract for HTK-1 trainer helicopters was awarded during September 1950. Work began immediately, and the first of two prototypes was delivered during April 1951. A total of 29 production machines followed with the last example completed by 1953.

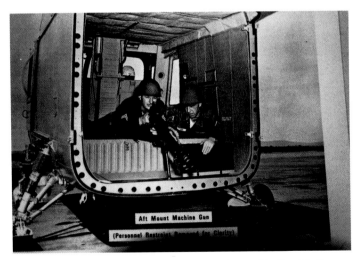

Though this .30 cal. mount was never adopted by the Air Force, a similar system using heavy weapons suspended from the rear fuselage saw limited use in Southeast Asia. (Kaman)

This aft view of the HH-43 shows the visibility afforded by the clam-shell doors, along with the bracing used between the tail booms. The post atop the elevator was a mount for the ADF antenna, which ran to the rotor pylon. (Dale Mutza)

First introduced by the Air Force during 1965 as part of the Huskie's rescue equipment, Kaman's forest penetrator became so successful that it saw widespread use in other services. Here, it is mounted aboard a U.S. Army UH-1H "Huey." (U.S. Army)

The turbine engine installation permitted full use of the fuselage internal area. The H-43B had twice the cubic capacity of its piston-powered predecessor. (National Archives)

The H-43B serial number 58-1846, along with number 1847, was evaluated by the Air Force Flight Test Center of the Air Research and Development Command at Edwards AFB during early 1960. The objective of the trials was to determine if the Huskie could meet operational requirements. (National Archives)

Though basically a training aircraft, the HTK-1, like the planned HOK-1 design, could easily be adapted to medical evacuation duty with an outward-opening canopy section to facilitate litter loading. Production HTK-1s (Kaman Model K-240) differed significantly from the prototypes in that they were three-seat models powered by a 235 hp 0-435-4 engine, compared to a 225 hp 0-435 installed in the two-seat prototypes. Although it was Kaman's first volume production helicopter, the HTK-1 became best known for achieving aviation milestones in the virgin territory of drones and turbine power.

Since no engines had been designed specifically for helicopters, reciprocating engines that powered airplanes were used. Their weight-to-horsepower ratio, an extremely critical factor in helicopter design, left rotary-wing aircraft severely underpowered. Some bulky engine configurations reduced a helicopter's payload to the point that the aircraft's usefulness was legitimately questioned. Once again, Kaman became intrigued with a concept that would put helicopters on somewhat of a par with their fixed-wing contemporaries. The firm's engineers showed a keen interest in the Boeing 502-2 gas turbine engine developed for Navy shipboard use. While not fuel efficient, the turbine could produce 200 horsepower, enough for Kaman to sell the Navy on a proposal for an experimental installation in a K-225 helicopter.

For the modification, the Navy sent back one of the first two K-225s it had purchased. When the hybrid was completed, it made its flight debut as the world's first gas turbine helicopter on 11 December 1951. The simplicity, efficiency and even quietness of the system heralded a tremendous breakthrough for the aviation industry.

Savoring their success and confident in their ability to take helicopters into the jet age, Kaman engineers and de-

Since rotor-to-tail strikes plagued the H-43, resulting in their grounding during the evaluation period, the test aircraft had rods attached to the left inboard and right outboard rudders to register rotor blade contact with tail surfaces. (National Archives)

The H-43B demonstrates its lifting ability during 1959. (Kaman)

During a demonstration of the H-43B at Stead AFB, a photographer inadvertently captured a blade-to-tail strike on film. The severed tip of the left inboard tail fin can be seen falling toward the firefighter. At the controls was William A. Lyell, Air Training Command's officer for the H-43, who mentioned the 38-knot wind probably had a part in the incident. (Kaman via Lyell collection)

signers modified a HTK-1 with two Boeing 502-2 engines, claiming another first. The world's earliest twin gas turbine helicopter was successfully flight tested during March 1954.

Meanwhile, work continued on the HOK-1, which was the first helicopter designed to military specifications. The first of two XHOK-1 prototypes took to the air on 21 April 1953. The advantage of a large enclosed cabin to accommodate litter patients was offset by excessive turbulence caused by the bulky fuselage. As structural and control problems were addressed, a problem arose with rotor tracking as a result of increased blade size. No sooner was that resolved with innovative inflight pilot tracking control when trouble with the engine was encountered. The original HOK-1 used a Continental R-975-40 450 hp engine, the same used to power the HUP-1 built by Piasecki for the Navy. The engine was a reworked tank powerplant, which proved unreliable and caused the loss of a number of HUP-1s at sea. Shortly thereafter, the switch was made to the Pratt & Whitney R-1340-48 Wasp engine, a stalwart in the industry, which developed 600 horsepower.

To determine the suitability of the HOK-1 for U.S. Army utility use, a Marine HOK-1 was put through its paces for six months by the Army Aviation Board beginning on 25 October 1956. After nearly 200 hours of thorough testing, the Army found the HOK-1 more than suitable, however, it did not offer sufficient advantages over existing types in the inventory to justify immediate procurement. Especially appealing to Army aviators were the rotor configuration and blade-tracking system. The final evaluation report recommended that the type be considered for future Army use. The report listed the Canadian Army as an interested agency, which asked to review the test results. The decision to evaluate the HOK-1 was based on the favorable impression made by the HTK-1, which the

A H-43B, showing the final tail configuration, makes a hoist pickup. The right side cabin door could be slid aft as one or two units. (Kaman)

One of the first H-43Bs performs a test flight following overhaul at the factory, which included camouflage paint. (Kaman)

Army had tested during summer 1953 at Fort Bragg, North Carolina.

A total of 81 HOK-1s were produced for the Marine Corps, with the final example delivered during December 1957. A follow-on order of 24 similar aircraft for Navy utility duty, designated HUK-1, came off the production line from 1957 to 1959. They differed from HOK-1s only in Navy-installed equipment.

Under the revised designation system implemented during 1962, the HOK-1 became the OH-43D, while the HUK-1 was re-designated the UH-34C and the HTK-1 became the TH-43E.

Kaman's Model K-600 would remain the standard throughout its line of H-43 helicopters. The company wasted no time in soliciting sales of the K-600 in the commercial market, however, the attempt proved unsuccessful.

Kaman's achievements, along with studies in other aspects of rotary-wing aviation, compelled the military to set their sights on turbine power. In 1954 the Army held a competition for a new turbine-powered helicopter. Lycoming, which went to work on helicopter turbines shortly after the K-225's epic flight, won the engine contract. Bell Helicopter's entry, the XH-40, was selected as the aircraft. Since Lycoming's new T53 engine was ready for flight tests before Bell's helicopter was completed, the military turned to Kaman. Under Air Force contract, on behalf of the Army, the turbine powerplant was installed in a HOK-1 during spring 1956. Flight testing began on 27 September. Bell's XH-40, which was later redesignated UH-1, the renowned "Huey," first flew in November 1956, two months after Kaman's HOK-1 flight.

With an eye again toward the commercial market, Kaman proposed two versions of its turbine-powered HOK-1: in addition to the military HOK-1, there was to be a twin turbine model (K-600-4) powered by two British Turmo 600 gas turbines. Since the engines were type certified and in production, Kaman announced that the combination could be readily certificated for commercial use. Little came of the proposal, however, and renewed emphasis was placed on military orders.

During 1956 Kaman's Model 600 scored high in an Air Force evaluation of existing helicopter designs suitable for a newly-established crash-rescue mission. Based on the HOK-1/HUK-1 and designated H-43, the type was first flown on 19 September 1958. An initial contract for 18 H-43As was fulfilled with the last aircraft delivered by mid 1959. An additional "A" model was built, which later became a YH-43B. In view of the success with mating a HOK-1 with a turbine engine, Kaman switched completely to turbine power, another first among helicopter manufacturers.

During February 1958 the Air Force placed an order for 20 H-43Bs, powered by the 825 hp Lycoming T53-L-1A gas turbine engine. Though the "B" model was of the same general configuration as its piston-powered predecessor, it was completely redesigned to improve performance and load-carrying ability. The first H-43B took to the air on 1 November 1958, followed by the second prototype, which was destroyed in a crash on 14 December. The first production H-43B, serial number 58-1841, first flew on 13 March 1959, and deliveries began during June. Production models were equipped with the 860 hp T53-L-1B engine, which allowed a maximum speed at sea level of 120 mph. The H-43B could accommodate two pilots plus six passengers, or a pilot, medical attendant and four litters.

The H-43B's unique tail section underwent a series of modifications to cope with the problem of blade-to-tail strikes. After several incidents where blades contacted vertical tail

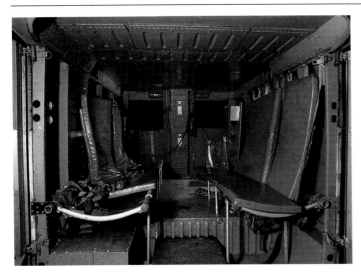

This view looking forward through the clam-shell doors of a HH-43F shows the cabin layout, which included a partial bulkhead behind the cockpit and cushioned passenger seats. (New England Aviation museum)

surfaces, the H-43 fleet was grounded on 4 May 1960. The grounding was lifted after operational aircraft were fitted with breakaway tail tips, and the aircraft were restricted to a very limited flight envelope. Finally, the empennage was vastly reworked to eliminate the problem.

The H-43B distinguished itself in the record-setting arena, breaking eight world records, four for Class E-1 helicopters (all gross weights) and four for Class E-l.d helicopters (3,858 to 6,614 pound gross weights). The first record was estab-

lished in FAI Class E-l.d on 9 December 1959, when a H-43B, piloted by two Air Force pilots, attained an altitude of 29,846 feet! The standard H-43B, in mission configuration, topped the previous record held by a Soviet MI-4 helicopter, which flew to 21,982 feet on 12 March 1959. On 25 May 1961 a H-43B reached an altitude of 26,369 feet with a 2,204-pound payload aboard, exceeding the previous mark of 24,491 feet set by a Soviet MI-4 on 26 March 1960. The type then went on to claim three time-to-climb records on 24 October 1961. The successive distances and times were 9,842 feet in 2 minutes, 41.5 seconds; 19,684 feet in 6 minutes, 49.3 seconds and 29,426 feet in 14 minutes, 30.7 seconds.

In the Class E-l.d category, a H-43B achieved an Altitude Without Payload record on 18 October 1961 by attaining an altitude of 32,840 feet. For Distance in a Closed Circuit, a H-43B traveled a distance of 655.64 miles on 13 June 1962. And for Distance in a Straight Line, a H- 43B traveled a distance of 888.44 miles.

A total of 202 H-43Bs were produced. A later model, born of necessity to cope with battlefield and hot climate conditions in Southeast Asia, was first flown in 1964. Designated the HH-43F, the improved version utilized the T53-L-11A turbine rated at 1,150 shp. The increased power allowed the "F" model to carry 11 passengers, plus crew. A total of 42 HH-43Fs were built, the last of which rolled off the production line during 1968. A total of 33 HH-43Bs were eventually brought up to "F" standards.

During its heyday, the H-43's safety record was the best ever established by a military aircraft. Its accident rate was far below the U.S. Air Force all-helicopter and fixed-wing rate.

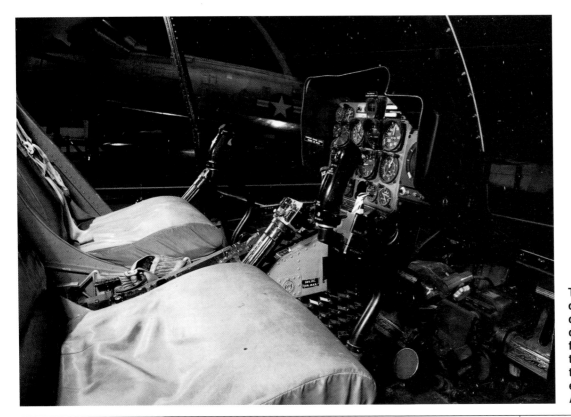

This view of the HH-43F's cockpit shows the practical controls layout. The cyclic control sticks are forward of the seats, while the collective/throttle control levers are to the left of each seat. (New England Air Museum)

**Unusual on this HH-43B are the split black and white markings and large national insignia applied over factory- fresh camouflage. The Huskie's complete serial number appears on the tail fin. (Kaman)**

The type had been in USAF service for nearly five years before it suffered its first fatal accident, which occurred at Stead Air Force Base.

Ironically, the unique characteristics and successive improvements, which ensured the success of the H-43 series, were also responsible for its demise. The dual rotor configuration was slow. In describing their HOK-1 helicopter, the Marines probably said it best, "The aircraft performed admirably—as long as the occupants were in no great hurry." After the Korean war, the military realized the need for speed in helicopters. The fact that dual rotors created significantly more drag than a single-rotor arrangement, and turbine engines became able to provide ample power to drive tail rotors,

spelled the end of the H-43 line. Even Kaman, though loyal to its original design, deviated from the dual intermeshing rotor configuration with its successful single-rotor/tail-rotor H-2 "Seasprite" series. Kaman's trademark servo flaps, however, were utilized on the Seasprite's conventional rotor.

After commercial operators found useful applications for surplus military H-43s in logging, petroleum exploration, agricultural spraying and firefighting, there became a strong demand for H-43s; so strong, in fact, that Kaman reopened a synchropter line in the 1990s. And so, the Huskie lives on in a new generation of helicopter, called the "K-Max," which too resembles a sparrow beating itself to death, but is able to perform like no other helicopter in the industry.

# Chapter 2
# The Aircraft

In 1959 a handful of U.S. Air Force pilots was selected to receive a factory checkout at Kaman's Bloomfield, Connecticut plant in a revolutionary new helicopter with a gas turbine engine. That was the good news. The bad news was that it came with intermeshing wooden blades. "Wood?," they thought. "Oh great!" They joked about eggbeaters with termite problems, but, in truth, were a bit apprehensive about flying a machine with vital parts made from a tree. Those fears were greatly alleviated when they toured the factory where the blades were manufactured.

It looked like an old folks home. Nearly everyone had gray hair and bifocals. It turned out that all those senior citizens were master craftsmen who had been working all their lives as cabinet and furniture makers. They were so skilled that they were using templates to work tolerances of less than a thousandth of an inch. The pilots were impressed, and relieved.

A few months later, the confidence of one of those pilots was boosted when his H-43B became trapped between a violent mountain thunderstorm and rock spires, while returning with minimum fuel from a rescue mission over Wyoming's Grand Tetons. The turbulence was the most violent he had experienced in a lifetime of flying. It cracked one of the spruce rotor blades from the trailing edge clear to the main spar, but everything held together. Had it not been for their great flexibility, the blades would certainly have failed.

Such accounts speak well not only for Kaman's workmanship, but his engineering and design abilities as well. The company would eventually deviate from all-wood blades in favor of metal blades, which could flex with more predictability.

The H-43 owes its success to two features which represent Kaman's contributions to rotary-wing aviation; the intermeshing rotor system and the servo flap. Charles Kaman

The Navy's first Kaman helicopter ordered into production was the HTK-1, all of which were delivered by 1953. Pictured here are Bureau Numbers 129302 and 128660. (Kaman)

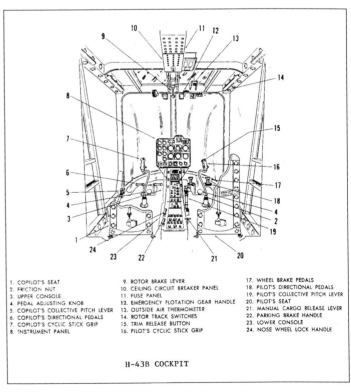

| 1. COPILOT'S SEAT | 9. ROTOR BRAKE LEVER | 17. WHEEL BRAKE PEDALS |
| 2. FRICTION NUT | 10. CEILING CIRCUIT BREAKER PANEL | 18. PILOT'S DIRECTIONAL PEDALS |
| 3. UPPER CONSOLE | 11. FUSE PANEL | 19. PILOT'S COLLECTIVE PITCH LEVER |
| 4. PEDAL ADJUSTING KNOB | 12. EMERGENCY FLOTATION GEAR HANDLE | 20. PILOT'S SEAT |
| 5. COPILOT'S COLLECTIVE PITCH LEVER | 13. OUTSIDE AIR THERMOMETER | 21. MANUAL CARGO RELEASE LEVER |
| 6. COPILOT'S DIRECTIONAL PEDALS | 14. ROTOR TRACK SWITCHES | 22. PARKING BRAKE HANDLE |
| 7. COPILOT'S CYCLIC STICK GRIP | 15. TRIM RELEASE BUTTON | 23. LOWER CONSOLE |
| 8. INSTRUMENT PANEL | 16. PILOT'S CYCLIC STICK GRIP | 24. NOSE WHEEL LOCK HANDLE |

H-43B COCKPIT

See Legend (OPPOSITE) for parts description

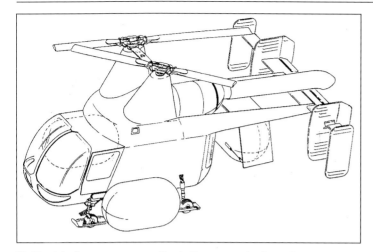

H-43 with the emergency floatation gear deployed.

Details of the H-43's intermeshing rotor hubs and canted pylons. A flat support brace between the pylons incorporated an anti-collision light. (National Archives)

invented the servo flap while working on Igor Sikorsky's VS-300 helicopter. Convinced that his concept could control rotor blade pitch better than conventional means, and receiving little support from his employer, United Aircraft's Hamilton Standard Division, he set out in 1945 to form his own company.

A helicopter rotor's direction and magnitude of lift is controlled by blade pitch. Changing the blade pitch of standard helicopters is accomplished by servo units and the mechanical movement of a "pitch horn" at the blade's root. Kaman's servo flap, which was attached to the blade's trailing edge and controlled by pilot input through push-pull rods, changed pitch aerodynamically, by actually "flying" the blade to the desired angle of attack, making the helicopter descend or climb.

Since the entire blade did not have to be twisted mechanically, the only control forces felt by the pilot, in the form of vibration, were those transmitted by the small flap. Not only did the servo flap stabilize the blade, rotor hub components were smaller and lighter and no hydraulic boost was needed.

The limited power produced by piston engines in early helicopters made any rotor design that minimized torque ap-

pealing. The anti-torque tail rotor used with a single main rotor design consumed about ten percent of precious power in a hover and nearly three times that amount during maneuvers. Kaman believed that two intermeshing rotors was simply more efficient. The "synchropter" design comprised two double-bladed rotors mounted side-by-side on separate pylons. Viewed from above, the rotors counter-rotated with the right rotor turning clockwise and the left rotor turning counterclockwise. Both rotors were mounted on splined shafts and synchronized through a single transmission. The counter-rotation completely nullified torque reactions, eliminating the need for a tail rotor.

Since the overlapping rotors operated in a greater volume of air, the aircraft was extremely stable, especially in a hover, and could easily outclimb the next generation of helicopters. The H-43's long blades formed a large disc area, which the Navy found to be a major drawback, citing deck-handling limitations. The H-43's low rotor disc loading allowed

| 1. Copilot's seat | 13. Outside Air Thermometer |
|---|---|
| 2. Friction Nut | 14. Rotor Track Switches |
| 3. Upper Console | 15. Trim Release Button |
| 4. Pedal Adjusting Knob | 16. Pilot's Cyclic Stick Grip |
| 5. Copilot's Collective Pitch Lever | 17. Wheel Brake Pedals |
| 6. Copilot's Directional Pedals | 18. Pilot's Directional Pedals |
| 7. Copilot's Cyclic Stick Grip | 19. Pilot's Collective Pitch Lever |
| 8. Instrument Panel | 20. Pilot's Seat |
| 9. Rotor Brake Lever | 21. Manual Cargo Release Lever |
| 10. Ceiling Circuit Breaker Panel | 22. Parking Brake Handle |
| 11. Fuse Panel | 23. Lower Console |
| 12. Emergency Flotation Gear Handle | 24. Nose Wheel Lock Handle |

The rescue hoist installation on the H-43A. The oil cooler filled much of the air intake opening at the base of the rotor pylons. (Lennart Lundh)

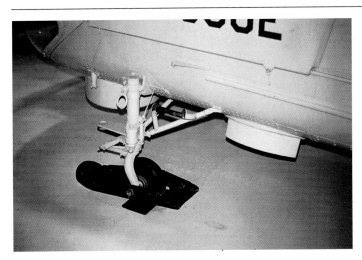

Details of the H-43A's forward landing gear. The tapered projection aft of the gear support is an ARC- 52 UHF antenna. (Lennart Lundh)

This view of the OH-43D (HOK-1) shows the bulged window in the sliding door. FM and HF antennas were added to the canopy framework. (Dale Mutza)

it to handle loads with ease. Besides having virtually no limitations, the unique rotor system permitted hovers equally well upwind or downwind, lift in rearward flight, and had the ability to start or stop in 70-knot winds.

Helicopters have their own peculiar way of addressing the old adage, "What goes up, must come down," called "autorotation." The term describes a helicopter's dimension of flight between loss of engine power and landing. The complex aerodynamic principles involved amount to good pilot technique and the rotor's ability to retain rotational energy, with relation to airflow. By being able to change blade pitch without engine power, the pilot is left with some options to control the descent. Most helicopter pilots consider autorotations simple maneuvers and prudent pilots practice them often. Pilots praised the H-43's ability to perform zero-airspeed auturotations and touch down gently as a feather. During the transition from powered flight to autorotation, the loss of con-

Visible through the OH-43D's windshield is the pilot's station on the aircraft's right and litter accommodations opposite. The copilot's seat and controls were removed and the canopy half swung open for litter access. (Dale Mutza)

The bulky installation of the Pratt & Whitney R-1340 piston engine in the HOK-1. The fuel filler neck protrudes from the cabin right side. (Dale Mutza)

This view of the HOK-1's cabin shows the collapsible litter apparatus fully extended. Tinted windows above the pilots' seats became standard on all successive models. (U.S. Army)

siderable altitude is a common occurrence, which a pilot can avoid only under certain conditions and with a delicate balance of power. Therefore, a flame-out at low altitudes and low speeds is especially dangerous.

One Air Force Huskie pilot experienced an engine failure at the dreaded combination of altitude and airspeed, from which, according to the flight manual, recovery was impossible. During a running takeoff, the pilot made it to a height of 50 feet and 50 knots when the engine quit. With expert use of the controls, he was able to turn the Huskie 110 degrees back to the ramp, radio a "Mayday" to the tower and tell the crew to hang on. He flared, rolled level and pulled the collective full up, putting maximum pitch on the blades, and touched down in a slight crab, forcing the right front nose gear to break off. To his credit, other H-43s had been destroyed after engine failure under more favorable conditions.

Directional control of the rotor system was accomplished by the cyclic system. Movement of the cyclic stick in any direction caused both rotor discs to tilt and fly the aircraft in that direction, and at a speed relative to the amount of stick movement. Moving the stick sideways caused the corresponding disc to tilt, with the opposite rotor following the movement only partially so its blades wouldn't strike the opposite hub. The H-43 could be turned by pushing foot (rudder) pedals, which changed the pitch relationship between the rotors.

Pilots claim the H-43 was very easy to fly, but also maintain that it seemed to have a mind of its own. Especially peculiar during power changes in early models was the aircraft's tendency to enter a phase where right rudder was required to turn left and vice versa. To compensate for the "control reversal," a manifold pressure system made the correction for the pilot. However, its failure provided some thrilling moments for those aboard.

Airplane-like stability, particularly in high-speed forward flight, was possible with tail booms that incorporated vertical and horizontal surfaces. Early single-boom designs gave way

The instrument panel and console of the HOK-1. The pilot's collective lever is to the immediate right of the console. (U.S. Army)

The H-43A's piston engine installation included a 7- gallon capacity oil tank. (Lennart Lundh)

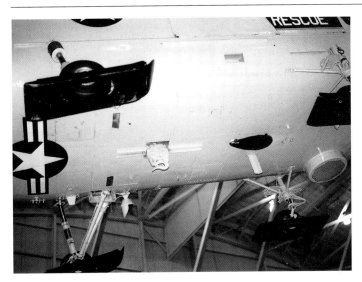

Details of the H-43B's undersides include a teardrop- shaped ADF loop antenna next to a UHF blade antenna, cargo hook and floodlight. The round flat fairing houses a UHF DF antenna. (Lennart Lundh)

This underside view of the H-43B's empennage shows the horizontal floating elevator surfaces to which the end plates were rigidly attached. (Lennart Lundh)

to twin booms, the tail surfaces of which underwent numerous changes throughout H-43 development.

Due to its rotor configuration, the turbine-powered H-43 was one of the most quiet helicopters in the world. In an effort to further reduce noise levels, Kaman conducted a two-part Quiet Helicopter Program under contract with the U.S. Army's Aviation Material Laboratories during 1969. Modifications to the rotor, drive system and engine of a HH-43B during the program resulted in significant noise reduction.

Technically, Kaman's original K-125 helicopter was a crude but successful craft of welded steel tube construction. The installation of a Lycoming six-cylinder, fan-cooled 0-435-A2 engine gave it a top speed of 72 mph. Following closely on its heels was the K-225 commercial model, two of which went to the U.S. Navy and one to the Coast Guard.

The K-225 featured an open cockpit with the pilot's seat up front and two passenger seats behind. The 0-435 engine installation gave it a cruising speed of 65 mph. Its vertical rate of climb was 300 feet per minute, and it had a ceiling of 12,000 feet. Its autorotative rate of descent was an impressive 1,250 feet per minute. Empty, the type weighed 1,800 pounds, and its gross weight was 3,000 pounds. Other features of the K-225 included a 32-gallon capacity gravity fuel system and hydraulic rotor brake.

The transformation from K-225 to HTK-1 trainer resulted in an entirely new design, retaining the unbeatable combination of intermeshing rotors and servo flaps. The prototype, which made its debut at the Helicopter Forum during April 1951, was basically a K-225 with an enclosed cockpit and covered fuselage, one major change being a raised tail boom and quadricycle landing gear. Horizontal and vertical tail surfaces were connected to the collective pitch control for greater stability.

After several major modifications, it emerged from production as an entirely new design. Since shortening of the tail boom caused a loss of moment arm, the oval fins were increased from 10.5 to 12.5 square feet. Ventral and dorsal fins, each 8.5 square feet, were added to the tail boom. After a number of fins were damaged during nose-high landings, an ingenious Navy metalsmith came up with a solution: wooden hunting bows purchased at local sporting goods stores were attached to the fins to act as tail skids. The prototype HTK-1 was equipped with K-225 rotor blades, which had equal chord throughout. Taking their place on production models were newly designed blades that were narrower, widened at the servo flap and tapered toward the tip. Unlike the two-seat prototype, the three-seat production HTK-1 used a Lycoming 0-435-4 engine which delivered 235 horsepower.

The "bear paw" skid landing gear and attachment fittings. Though Kaman introduced the gear during 1956, few appeared on Marine HOK-Is. (U.S. Army)

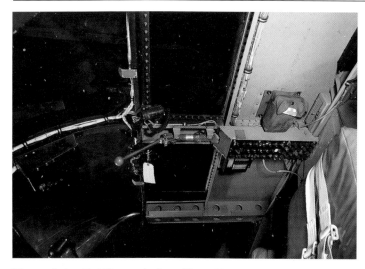

View of the H-43's cockpit ceiling. The long-handled device between the green-tinted windows is the rotor brake control. (Kaman)

The variant could be adapted to medical evacuation duty with provisions for a litter, loaded through the outward-swinging left half of the canopy.

The most prolific of the series, Kaman's Model K-600, began with the HOK-1. Ordered into production during 1950 for the U.S. Marine Corps, the HOK-1 was a vast deviation from the HTK-1. Structurally, it incorporated twin tail booms, which not only provided stability, but made room for the engine installation. After Marine trials, a large vertical stabilizer was added to the center tail section. Its 600 hp Pratt & Whitney, nine-cylinder R-1340-48 radial engine was mounted at the rear of the fuselage at a 35 degree angle to allow direct drive to the rotor gear box. The engine was fan-cooled through a shutter-controlled air inlet between the rotor pylons. Large rear fuselage doors (which were seldom used operationally except in cold climates) provided access to the powerplant. A single transmission drove the 47-foot diameter rotors, which were standard for the K-600 series. The blades could be positioned parallel to the fuselage for storage.

With a capacity for five persons, the HOK-1 was used for cargo hauling, search and rescue, observation and medical evacuation. As an air ambulance, it could accommodate two stretcher patients and an attendant, in addition to the pilot, who occupied the right seat. Stretcher loading was through the hinged left plexiglass nose. A cargo hook structure could be attached externally to the belly, and an electrically-operated hoist, designed to lift 600 pounds, could be mounted over either side of the cabin. Large sliding doors, which incorporated bulged windows, were on each side of the cabin.

The HOK-1's semi-monocoque aluminum fuselage was 25 feet in length, 11 feet, 9 inches in height and 7 feet in width (excluding rotors and landing gear). Its overall length was 47 feet with a height of 15 feet, 6 inches. The aircraft's empty weight was 4,490 pounds with a maximum gross weight of 7,500 pounds. The non-retractable landing gear comprised two main rear wheels with hydraulic brakes and two swiveling, self-centering forward auxiliary wheels. During May 1956, Kaman made available "Bear Paw" skis which attached to each wheel to prevent the aircraft from sinking into soft surfaces. Marine HOK-1s seldom, if ever, used the gear, and float landing gear made an even smaller impression.

At a normal loaded weight of 5,800 pounds, the HOK-1 had a maximum speed of 104 mph at sea level. It cruised at 58 mph and climbed at a maximum rate of 1,300 feet per minute. Its service ceiling was 18,000 feet, and a 102-gallon fuel capacity gave it a range of 180 miles.

The Marines found the HOK-1 to have excellent control and maneuverability through all weight ranges. They claimed it outperformed all other types in a hover, and praised its visibility from the cockpit and ease of maintenance. The HOK-1's major drawback, they noted, was its lack of speed.

The H-43A for the Air Force was based directly on the Marine HOK-1 and Navy HUK-1, differing only in the addition of rescue and firefighting equipment and tail configuration. A

Seen here in the HH-43F cockpit are the pilots' collective levers, which incorporate throttles, and cyclic control sticks. (New England Air Museum)

major change in the H-43A's tail arrangement improved stability and directional control during autorotation. The elevator surface was widened by two feet, giving the tail an overall width of nine feet. Two inboard ventral and dorsal fins replaced the massive center fin familiar to the HOK-1. The H-43A's servo flaps were enlarged slightly over those of earlier models.

Like the Marines, the Air Force found their version of the Model K-600 underpowered. When carrying a 1,000-pound fire suppression kit in hot climates, H-43A crews often reduced the fuel load to 200 pounds, giving an endurance of only 20 minutes. Especially appealing to the Air Force was the ability to transport a complete H-43A in a Fairchild C-119 cargo transport and two in a Douglas C-124. The first two H-43As were temporarily retained by Kaman for pilot and mechanic training, while the first delivery was made with the third machine in November 1958.

The switch to turbine power resulted in the H-43B, which exceeded Air Force requirements for the crash-rescue mission. While similar to the H-43A and designed for the same type of missions, the H-43B represented a radical change in fuselage configuration and performance. The turbine engine claimed a number of advantages over reciprocating types. Most noteworthy were noise and vibration reductions, less

The H-43's emergency flotation gear canisters were mounted beneath the cabin door openings. The actuating gas cylinders were immediately forward of the fairings. (U.S. Air Force)

moving components, increased power and better payload arrangement. Installing the smaller, lighter turbine above the H-43B's fuselage nearly doubled the cubic capacity of earlier models. Not only was the entire rear of the aircraft freed for cargo, but clam-shell type doors, which formed the rear of the cabin, were added for easier movement. The spacious cabin measured just over five feet in width, four feet in height and nearly twelve feet in length.

The new fuselage could accommodate seven passengers, or four litters and attendant, plus pilot. A partial bulkhead separated the cabin from the cockpit. To increase rescue efficiency, the rescue hoist was mounted on the same side as the pilot, above the right side cabin doorway. The distance between rotor hubs was increased 25 percent over the H-43A to reduce the blade coning angle and increase turn moments.

The powerplant originally selected for the YH-43B prototype was Lycoming's T53-L-1A, which was quickly upgraded to the T53-L-1B in production. A short upturned exhaust pipe was soon replaced by a long pipe that extended rearward over the tail's horizontal stablilizer. The T53-L-1B produced 860 shp, which was derated to 825 shp. The turbine installation gave the H-43B a top speed of 120 mph at sea level and a cruise speed of 110 mph. A 2,000 feet per minute rate of climb was possible, along with an incredible 25,700-foot service ceiling, fully loaded.

To improve the H-43B's hillside landing characteristics, the four-wheel gear track was widened to 8 feet, 4 inches and the gear itself was stiffened. Air Force pilots usually locked the self-castoring nose wheels since ground taxiing required a touchy combination of minimum engine rpm and high blade pitch for control. Since a flight control reverser caused problems with lesser pitch, it was easier to hover-taxi.

The H-43B's maximum gross weight was 7,500 pounds with an empty weight of 4,200 pounds. The fuel capacity of its predecessor was nearly doubled to 198 gallons, giving it a range at cruise speed of 185 nautical miles.

Provisions for the installation of emergency flotation gear were introduced on the H-43B. The system consisted of three inflatable gas bags folded into fairings at each side of the fuselage and between the tail booms. The bags were inflated by electrically-triggered gas generator units. The gear underwent its first operational testing at Thule AB, Greenland, during late 1963 by Detachment 1, 54th Air Rescue Squadron. Crews discovered a glaring drawback to the floats: once activated, the pressurized bags pressed against cabin doors, making them very difficult to open.

The original tail section of the H-43B comprised a "floating" horizontal elevator with fixed end fins, which acted as rudders, and two ventral fins in the center. The ventral fins were replaced by an identical pair atop the elevator. A problem with the arrangement was quickly remedied by an engineering change—or so it was thought. In a series of blade-to-tail interference incidents, aft cyclic travel had caused the rotor blades to hit the two inboard vertical fins. Three blade-

**The HH-43B serial number 58-1853 in fresh camouflage paint. The Huskie's two-piece right cabin door could be slid aft as a unit or individually at the cabin or pilot's position. (Kaman)**

to-tail strikes occurred at the Kaman facility between September 1959 and March 1960. Two more occurred during April 1960 at Stead AFB, one of which was a demonstration flight for officials. Kaman recommended that the Air Force impose minimum rpm flight restrictions and installed a spring device to warn pilots of excessive aft cyclic. The situation was further aggravated on 4 May 1960 when an over-zealous pilot applied aft cyclic with the predictable results.

That day all H-43Bs were grounded and an evaluation of the problem was soon under way. Aircraft serial number 58-1849, equipped with the original empennage, and number 59-1548, sporting a modified configuration, were tested at the Air Force Flight test Center at Edwards AFB.

To provide the necessary blade-to-tail clearance, the inboard dorsal fins were removed and the vertical stabilizers were lowered 14 inches. Additional vertical stabilizers were added outboard of the original rudders on 20-inch elevator extensions to maintain directional control qualities. As added protection, the upper and lower 10 inches of each vertical surface was constructed of fiberglass in the rare event of a blade strike or tail-low landing. In addition, the drooped edge of the elevator (for delaying stall) was eliminated, and a centering support bungee was attached between the tail booms and elevator. Retrofit kits were delivered for field installation, and H-43Bs after serial number 59-1568 were produced with the new tail.

Extensive modifications of the H-43B (re-designated HH-43B in 1962) resulted in the armored HH-43F, which had a top speed of 120 mph. Equipped with the T53-L-11A, rated at 1,150 shp, the "F" model fulfilled the need for an improved variant in Southeast Asia's hot climate and combat environment. The normal fuel capacity was increased to 350 gallons, which allowed a range of just over 500 miles at a takeoff weight of 8,270 pounds. The HH-43F's empty weight was 4,620 pounds, with a maximum gross weight of 9,150 pounds. Its maximum rate of climb was 1,800 feet per minute, and its service ceiling was 23,000 feet.

# Chapter 3
# Approval from the Navy

The HH-43 line of helicopters had its humble beginnings in the U.S. Navy. Though the type was instrumental in pioneering the use of helicopters for the Marine Corps and shipboard duty, it would not reach the status of its Air Force cousin, though moderate compared to jets of the period. In a Navy where emphasis was placed on the development of ships and fixed-wing aircraft, acceptance of the helicopter remained cool. Part of the problem stemmed from the common perception in aviation communities that anything labeled "utility" was substandard, which seemed to apply particularly to helicopters. Few personnel and funds were devoted to the helicopter, which started out behind and lagged considerably throughout its first decade of naval service. Despite established Fleet policy, ASW helicopters were often launched in place of utility types, whose meager electronics paled by comparison.

During late 1949, the U.S. Navy bought a Kaman K-225 for evaluation. Another was sold to the Coast Guard and a third, which Charles Kaman convinced the Navy they needed as a backup, was also agreed upon. Those three orders averted financial ruin for Kaman and put the company on firm footing. During March 1950, the first machine (BuNo 125446) was flown to NAS Patuxent River, New Jersey, for demonstrations and training of the Navy pilots who would fly the aircraft.

Although Sikorsky and Bell proposals had been entered in the competition for a Marine observation helicopter, successful evaluations of Kaman's K-225 encouraged the Navy to consider the type. Sikorsky's S-52 won the competition, however, Kaman's K-225 was selected as a backup design. After two XHOK-1 prototypes were manufactured and put through their paces, four HOK-1s were delivered to the Marine Corps during mid 1950. When war broke out in Korea, Congress granted large appropriations to the Navy, which decided to buy more helicopters. Luckily for Kaman, that meant orders for 77 additional HOK-1s. Kaman then pressed for a trainer variant to narrow the gap between delivery of the original K-225 and the HOK-1. The first example was completed during April 1951 and sent to the Naval Air Test Center (NATC) at NAS Patuxent River for evaluation. A total of 29 HTK-1s were built and delivered to the Navy between 1951 and 1953. One machine was later acquired for drone conversion, while another became the first twin turbine-powered helicopter.

Before resigning them to their intended role as trainers, the Navy used the HTK-1s all over the world as utility helicopters beginning in 1952. Operations that ranged from icebreaker duty to observation flights in the Philippines resulted in several production refinements. When the HTK-1s finally assumed their training mode, they were assigned to Helicopter Training Unit One (HTU-1) at NAAS Ellyson Field, Florida, which was part of the Pensacola Naval Air Training Command. There, the HTK-1 took its place among Bell's HTL and Piasecki's HUP helicopters. After initial training in the HTL,

The first of two XHOK-1 prototypes (Kaman's Model K-225) during Flight Test trials at the Naval Air Test Center. (MAP)

With a fully enclosed airframe, the HTK-1 bore little resemblance to the K-225, from which it evolved. (Kaman)

The first production HOK-1 leads a flight of Kaman helicopters: a float-equipped HTK-1 and two K-225s. (U.S. Naval Institute)

Painted overall Orange-Yellow, this HTK-1 wears the markings of Helicopter Training Unit One (HTU-1). Kaman's trademark servo flaps are clearly visible on the rotors. A locally-fabricated skid has been added to the center tail fin. (Balogh via Menard)

Marine students flew 30 hours in the HTK-1, while fledgling Navy and Coast Guard fliers went on to the HUP.

Ironically, the HTK-1 was ill-suited for training Navy and Marine helicopter pilots. Being too easy to fly, it didn't prepare pilots to operate the more common tail rotor-equipped helicopters. With Kaman's less complicated, torque-balanced machines, the pilot had only to come up on the collective (which provided lift) and not worry about tail or cyclic stick problems. Charles Kaman reveled in the phrase, "Just give me a screwdriver and a pair of pliers, and I'll foul up the rigging enough to make it feel like a tail rotor configuration." Though he was correct, it didn't win him any favors in some circles.

The second production HTK-1 practices water landings during float gear evaluations. (Kaman)

Seen here in unusual markings while assigned to the NATC for Flight Test, this HUK-1 served as the prototype for the series. (Steve Miller)

This HUK-1, along with Bureau Number 146306, was assigned to NAS Barbers Point, Hawaii as station aircraft, beginning in late 1958. (R.A. Carlisle via NMNA)

The HOK-1 made its first flight test on 21 April 1953. The Navy also took delivery of 24 utility version HUK-1 helicopters, all of which were delivered by 1958. After initial trials at NAS Patuxent River during late 1957, the first prototype YHUK-1 went aboard the carrier USS WASP in the Atlantic for the formal Fleet Introduction Program (FIP) testing. The HTK-1 was phased out of the training syllabus during 1957, and surviving examples became TH-43Es during 1962.

To manage worlwide helicopter utility operations, the U.S. Navy had established units on opposite coasts on 1 April 1948. Helicopter Utility Squadron One (HU-1) began operations at California's Ream Field to serve the Pacific Ocean, while HU-2 set up shop at NAS Lakehurst in New Jersey to cover the Atlantic and Mediterranean. Detachments within both units ensured dispersion of aircraft throughout their respective regions.

Few assignments in Navy aviation could match the variety and interest of those drawn by both squadrons. In Fleet service, HUK-1s and HTK-1s were based aboard a variety of ships that included carriers, anti-submarine ships, cruisers, icebreakers, LSTs and hydrographic survey vessels. HU-2 tested its HTK-1s aboard icebreakers during the mid 1950s, while HU-1 hauled equipment across the arctic during 1955 for construction of the DEW line. Hydrographic teams and instruments were flown to remote mountain regions, and helicopters became the vital link between ships in a task force, hauling personnel and cargo.

While not the largest mission of utility sqaudrons, the most vital was that of "plane guard," which involved rescuing airmen from open seas. In a role not duplicated by other rescue agencies, Navy and Marine helicopters took up a position near a carrier during aircraft launch and recovery operations.

A HUK-1 assigned to NAS Jacksonville, Florida during 1961. (Leo J. Kohn)

The fairing attached to this HUK-I's fuselage housed AN/ARC-12 radio equipment. (MAP)

An HUK-1 of Helicopter Utility Squadron Two (HU-2) during carrier evaluations aboard the USS LEYTE (CVS-32) during August 1958. Suitability trials for the HUK-1 were conducted aboard the USS WASP (CVS-18). (U.S. Navy)

The HUK-1 during cargo evaluations at McGuire AFB, New Jersey. (U.S. Air Force)

Helicopter Utility Squadron Two

Helicopter Utility Squadron Two

A HUK-1 performs a water rescue demonstration at NAS Lakehurst, New Jersey during November 1958. (U.S. Navy)

Personnel of HU-1 "Flying Angels" test a 20-foot nylon rescue ladder they devised for their HTK-1 during July 1958. The HTK-1 shared rescue duties at NAS Barbers Point with a hoist-equipped Sikorsky H03S-1. (U.S. Navy)

Helicopter Utility Squadron One

In the event an aircraft ditched or a pilot ejected, the helicopter quickly moved in to perform the rescue. Rescue swimmers aboard the chopper provided a quick response to aircrew who were injured, trapped in their aircraft or entangled in parachute shrouds. Prior to the helicopter's arrival, a destroyer was pulled out of the task force screen for plane guard duty. The unique practice continues during modern carrier operations.

Squadron commanders used their firsthand experience to work closely with the Navy Air Material Center and Aircrew Equipment Laboratory to develop equipment and techniques to ensure rapid and safe aircrew recovery—a challenge, considering the steady advancement of carrier aircraft. The commanding officer of HU-2 made enough noise about the lack of adequate rescue equipment and techniques that the Chief of Naval Operations added research and development to the unit's mission statement, which was a first for a Fleet squadron.

Credit for the first cable-lowered rescue seat concept and development goes to HU-2, whose members fabricated an inverted "Tee" with opposing seats. A flotation pack was added, and follow-up versions included a third seat and aluminum

The HUK-1 proved ideal for shipboard delivery of personnel and supplies. (U.S. Navy)

Except for internal equipment, the HUK-1 was identical to the Marine HOK-1. Here, the second-built HUK-1 has an external cargo hook beam attached to its underside. (Kaman)

Navy personnel at the Kaman plant on 1 August 1958 to take delivery of the HUK-1. Kaman test pilot Bill Murray is at far left. (Kaman)

Credit for the H-43's first rescue seat design goes to HU-2, which constructed this interim version during 1957. (U.S. Navy)

Helicopter Utility Squadron Two also developed this experimental "scoop" net during 1959 as part of its involvement with devising rescue techniques and equipment. (U.S. Navy)

flotation tanks. Other interested agencies were kept informed of the Navy's activities involving search and rescue equipment and techniques. Foremost among them were USAF SAR headquarters, Navy Development Squadron One (VX-1) at NAS Patuxent River, and the Naval Aviation Safety Center at NAS Norfolk.

Despite its favorable record, Kaman's design was deemed unsuitable for shipboard operations. Rotor tips whirled dangerously close to decks, and the aircraft was considered cumbersome for shipboard operations. Rumblings from some Navy pilots were heard concerning excess vibration and lack of maneuverability. One of the many dangers faced by aviators flying early helicopters was shipboard and ship-vicinity turbulence, which affected the aircraft's control limits.

Kaman's design also had two brief contacts with the U.S. Army. The first occurred during production of the HTK-1, one of which underwent suitability trials by Army ground forces at Fort Bragg, North Carolina, during late 1952. Little came of the study, and four years would pass before the type was again considered for "Army green." During the post-Korean war period, the Army began formulating plans for a helicopter

force comprising three types of helicopters: a small observation type, a medium-size utility type, and a large transport. A HOK-1 was borrowed from the U.S. Marine Corps during 1956 for thorough evaluations by several test board agencies at Fort Rucker, Alabama. Though Kaman's entry passed muster, the Army revised its requirements for Utility Helicopter Characteristics during 1959, which favored the new sleek Bell "Huey."

Likewise, Navy HUK-1s were phased out in lieu of the Huey and Kaman's turbine-powered UH-2—both were tail rotor configurations. HUK-1s were re-designated UH-43Cs in 1962 and remained in service until 1964. Despite the few shortcomings the Navy HTK-1 and HUK-1 possessed, their unique and positive features placed them at the forefront of naval helicopter development.

Like most U.S. Naval aircraft of the period, early Kaman helicopters were painted overall glossy Sea Blue. Later directives specified a change to glossy Engine Gray, which gave way to Light Gull Gray. Glossy Orange Yellow was applied to trainer helicopters beginning in late 1952 due to its high visibility against terrain during low altitude operations.

# Chapter 4
# The Marines Become Interested

United States Marine Corps interest in rotary-wing aviation dates back to 1932, when a Pitcairn autogyro was evaluated to determine its potential for military use. Although the oddity was found to have very limited capabilities, especially with regard to payload, trials with a similar type were conducted three years later. After Igor Sikorsky flew the VS-300 in 1939, the three major helicopter manufacturers (Sikorsky, Bell and Piasecki) began turning out successful rotary-wing machines, which directly influenced the development of Navy and Marine helicopter programs.

The Marine Corps took its cue directly from the Navy, which had formed its first operational helicopter unit on the heels of a helicopter development program initiated on 16 May 1946. Helicopter Development Squadron Three (VX-3) was officially established on 1 July at New York's Floyd Bennett Field. Before VX-3 was decommissioned in early 1948, its training program turned out the first Marine helicopter pilots.

The HOK-1 production line at Kaman's Bloomfield, Connecticut plant in 1954. (Kaman)

HOK-1s await delivery to Marine Corps units. (National Archives)

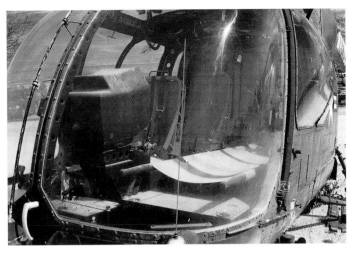

The copilot's seat and controls could be removed from the HOK-1's cockpit to accommodate two litters, which were loaded through the hinged windshield. A medical attendant occupied the seat behind the pilot. (Dale Mutza)

The color Marine Field Green was specified for HOK-1s during the 1960s. The cable drum for the rescue hoist was located between the rotor pylons. (Dale Mutza)

A HOK-1, assigned to MCAS El Toro, California, lowers its rescue seat to a victim at Newport Beach during December 1959. The aircraft was painted overall Orange-Yellow. (Kaman)

The Marine helicopter program got its official start with the establishment of Marine Helicopter Squadron One (HMX-1) on 1 December 1947 at Quantico, Virginia. Marine headquarters specified that the unit's two-fold mission was to develop tactics and techniques related to troop movement in amphibious operations, and to evaluate a small helicopter to replace the OY observation airplanes used for gunfire spotting, reconnaissance and liaison missions.

HMX-1 operated three primary helicopter types when, in fall 1949, the Kaman Aircraft Corporation announced its desire to show off its new Model K-190. Billing it as an observation helicopter had the desired effect by catching the attention of Marine aviation commanders. Adding to its appeal were advanced and unique features not found in the helicopters of HMX-1. Following initial evaluation of a single K-190 by the Navy Bureau of Aeronautics, the aircraft was turned over to HMX-1 during late December 1949. So impressed was the bureau with the helicopter's servo system alone, that a recommendation was made to procure another. However, that recommendation would be put on hold while the Marine Corps busied itself with the war in Korea and its quest for an ideal assault helicopter.

A HOK-1 at Dayton, Ohio during 1954. The engine compartment doors were rarely used except in cold weather. Note the significant forward tilt of the tail fins. (Balogh via Menard)

A HOK-1 of Marine Helicopter Development Squadron One at MCAS Quantico, Virginia during November 1962. (Lucabaugh via Steve Miller)

It was 1952 before the Marines once again focused on procurement of utility/observation helicopters. Not only were the Marines in desperate need of a utility aircraft, but the demand for observation helicopters was based on plans to replace all OY fixed-wing aircraft in observation squadrons (VMOs) with helicopters. Mindful of the favorable impression left by Kaman's K-190, the Marines ordered a second machine. During the period since the K-190's debut, the type had evolved into the improved K-225. After the first machine was brought up to K-225 standards, both were designated XHOK-1s and carried U.S.N. bureau numbers 125477 and 125478. Evaluation of the K-225 observation helicopter concluded that the design possessed superior flight characteristics, particularly stability, control and ease of handling. Those attributes, coupled with the low performance ratings of

Sikorsky's H05S-1 acquired during 1951, prompted a Marine order for 46 K-225s to eventually replace the H05S-1. After a number of radical design changes, the production version of the K-225 became the HOK-1 (Kaman Model K-3 and K-600), the first of which was delivered during April 1953. Though the Marine HOK-1 order was a first, it was not the only Kaman helicopter in the Marine inventory; the HOK-1 had been augmented by four HTK-1 variants assigned to Marine aviation for evaluation, one of which remained in the inventory until 1958. HOK-1 deliveries were completed by December 1957.

Immediately after its introduction to the Marine Corps, the HOK-1 was put through its paces during Bureau of Inspection and Survey trials and a Fleet Introduction Program (BIS/FIP) at Quantico and NAS Patuxent River. A number of problems were discovered, some of which resulted in two

Army personnel practice loading stretcher patients aboard a Marine HOK-1 at Fort Rucker, Alabama. (U.S. Army)

A HOK-1 assigned to Marine Observation Squadron Six during February 1961. Painting directives during that period called for prominent warnings about the rotor design's proximity to the ground. (John Hopton)

A HOK-1 of HMX-1. The unit's "XM" code was changed to "MX" during 1959. (National Archives)

After a fact-finding trip to Algeria in 1957 to observe armed French helicopters, the Marines conducted tests during early 1959 with French-designed SS-11 wire-guided air-to-surface missiles on a HOK-1. Though such tests proved highly successful, the dilemma of arming Marine helicopters would remain well into the 1960s. (U.S. Army)

This HOK-1 features a modified center tail fin and external fairings for radio equipment. (U.S. Marine Corps)

This view illustrates how the FM antenna was attached to the center windshield frame in such a way that it followed the windshield's contour. (U.S. Marine Corps)

A HOK-1 of VMO-2 participates in Operation Seahawk at Pohang, Korea during 1960. (National Archives)

Injured Marines are loaded aboard a VMO-6 HOK-1. (U.S. Marine Corps)

**A HOK-1 of VMO-1 makes a practice pickup with an experimental rescue net. (U.S. Naval Institute)**

fatal accidents. Once the problems were corrected and Marine aviators overcame their reluctance to fly the craft, the HOK-1 obtained acceptance from the Marines, especially in view of its remarkable stability. With its front section constructed mainly of plexiglas, the HOK-1 excelled as an observation platform. Equally impressive was its adaptability to the medevac role, thanks to a configuration that allowed two stretchers to be loaded through a hinged windscreen.

In observation squadrons, HOK-1s were teamed up with Cessna OE-1 (later O-1B) and OE-2 "Bird Dog" fixed-wing aircraft to serve as the eyes of the Fleet Marine Force in the Atlantic and Pacific. Throughout the 1950s and '60s, the Marines operated three observation squadrons: VMO-1 at MCAF (Marine Corps Air Facility) New River, North Carolina; VMO-2 at MCAF Futema, Okinawa, and VMO-6 at MCAF Pendelton, California. Each was initially equipped with 8 air-

craft, which soon increased to 12. The remainder were dispersed among air stations as rescue aircraft and HMX-1, which operated four. In the hands of HMX-1 crews, the HOK-1s underwent tests with various systems, including the French SS-11 wire-guided missile. The Marine HOK-1 inventory peaked during 1958 with 45 of the type on hand.

During 1960, the early concept of all-helicopter VMOs gave way to a mix of 12 fixed-wings and 12 helicopters per squadron. Most unusual, especially by today's specialization standards, was the fact that many observation squadron pilots were qualified in both types and interchangably flew HOKs or Bird Dogs. The HOK performed admirably throughout its operational career, flying missions that ranged from cargo hauling to dramatic rescues. When hurricane Diane ravaged the state of Connecticut during 1955, the HOK-1 gained notoriety when a photograph depicting a daring rescue with a HOK appeared in newspapers nationwide.

The HOK-1 became the OH-43D under the revised designation system during September 1962. By that time, the type had become obsolete, however, 35 remained in front line service. During late 1963, a crew from VMO-1 used their OH-43D to rescue a pilot and crewman of a light observation aircraft, which crashed in the ocean near the USS OKINAWA during a landing approach. The OH-43D landed on the ship to drop its cargo load and pick up two crewmen. They then flew to the crash site, hoisted the pair from their life raft and took them aboard the ship for medical treatment.

Several days later, another OH-43D, flying from St. Thomas to NS Roosevelt Roads intercepted a radio message between a Coast Guard HU-16 "Albatross" amphibian and the station tower. A person had been spotted in a sinking, water-filled boat, so the OH-43D crew joined in the rescue operation. Guided by smoke flares dropped from the Albatross, the OH-43D crew located the boat and hoisted the survivor from shark-infested waters. Not only were such actions taken in stride by Marine OH-43D crews, but many types of missions were flown that could not be performed economically by any other helicopter.

The three-year search for a single replacement for both OH-43D and OE fixed-wing aircraft had met with little suc-

cess. Finally, after experiments with Sikorsky's H-34 in the observation role proved futile, a suitable replacement was found. Ironically, though Kaman had accomplished much of the pioneering of turbine-powered helicopters, the OH-43D was replaced by Bell Helicopter's turbine-powered triumph, the UH-1 "Iroquois," best known as the "Huey." It wasn't until May 1965 that the OH-43D finally disappeared from the Marine inventory. A total of 81 HOK-1s had been procured for the Marine Corps. Long after its passing, tales of the unique OH-43D's accomplishments lingered where Marine helicopter pilots gathered.

Guidelines for Marine HOK-1/OH-43D paint schemes and markings followed those issued for Navy helicopters. The glossy Sea Blue livery that prevailed during the post-World War II years was replaced by overall glossy Orange Yellow after the Korean war. Although Navy directives called for a change to flat light Gull Gray, which was changed to glossy Engine Gray during July 1957, no Marine HOK-1s are known to have worn the former. Since 1952 Marine aviation had requested special consideration from the Navy Bureau of Aeronautics for a helicopter camouflage scheme. It wasn't until the early 1960s that Field Green was specified for Marine helicopters.

# Chapter 5
# Drones

Drones are basically unmanned craft that can perform missions without endangering the lives of crew, or that require the decision-making capability of a pilot. From the days of the autogyro, continued interest in pilotless aircraft resulted in various research programs involving drones.

Kaman's reputation for innovation and diversity ensured it a place at the forefront of such technological advances. Mindful of those attributes, the U.S. Navy relied heavily upon Kaman for technological pursuits. Like Lockheed's famed "Skunk Works," which referred to top secret programs, Kaman Aircraft used the term to identify a select team of specialists assigned to "sensitive" projects.

Charles Kaman developed a particular interest in anti-submarine warfare, which, during the post-World War II years, grew to be a significant factor in national defense. The threat of deeply submerged submarines with missiles warranted a fresh approach to a detection and attack capability. Kaman concluded that simple rotary-wing drones with listening sonars could be positioned, and moved, ahead of a naval convoy or task force. The concept would evolve into the Navy's Drone Anti-Submarine Helicopter (DASH) program.

Charles Kaman presented his ideas to the Office of Naval Research (ONR), convincing them that the drone, being unmanned, could provide long endurance at low gross weight. In theory, the conventional helicopter's many features designed around human safety could be eliminated and replaced by a robot system. Plans called for Kaman's patented rotor system, controlled by servo flaps, to provide a steady platform for an automatic stabilization system.

The ONR, after having first balked at the idea, reconsidered and came across with the funds necessary to produce a drone helicopter. Since the ONR dealt specifically with the theoretical aspect, the Bureau of Aeronautics (BuAer), which

The first Kaman drone helicopter, the HTK-1K nicknamed the "Yellow Peril", in its original configuration, is tethered for preliminary flight trials during Spring 1953. (Chuck Ellis via NMNA)

The "Yellow Peril," after modifications, is put through its paces during the late 1950s. (National Archives)

During 1958, one of Kaman's drones was tested with an electric motor in place of its piston engine. Current from a generator on the ground powered the motor. Figures on the aircraft's tail boom read, "NAVY KAMAN 137835." (National Archives)

During its first five years of operation, the QH-43G carried out highly classified trials as a high altitude antenna support vehicle operating from Navy destroyers. This was the first photo of the QH-43G released by Kaman, which depicts operations aboard the USS WRIGHT. (Kaman)

dealt with the actual hardware, was encouraged to sponsor the effort as well.

The HTK-1 was selected as the airframe for experimental work. When electronics were merged with its control system, pilots found that they could fly the craft hands-off. That breakthrough had anxious engineers wondering if a helicopter could fly without a pilot—an unthinkable feat during the early 1950s.

Their questions were answered in 1952, when Kaman launched the first of many drone programs, using an HTK-1 dubbed the "Yellow Peril," which had a corresponding pilot station on the ground. Its first flight was accomplished during May 1953, and by February 1955, had flown more than 100 hours. All helicopter flight phases, including sideways and backward flying and hovering, were performed, and all were made with a safety pilot aboard and within visual range of the ground operator. By April 1956, the program had progressed to the point where flights were performed by many, among whom were Navy officials, who, as old "sea dogs," had never flown anything in their life.

Kaman had convinced the Navy to pursue procurement, which led to an industry-wide competition that focused on the Kaman and Gyrodyne firms, the latter of which had built a

much smaller craft. Though the Navy chose Gyrodyne and the DASH program eventually failed, it did not spell the end of Kaman drones.

Army interest in developing a remote control reconnaissance capability, plus success with the Yellow Peril, led to a joint Navy-Army contract for three drone helicopters in June 1955. Kaman teamed up with RCA to mount a television camera in the nose of one of the modified aircraft, which were designated HTK-1Ks. It was then flown to Washington, D.C., where Army officials in a tent, with no view of the helicopter, remotely flew the machine along railroad tracks to Alexandria, Virginia. The Marines, who were interested in the drone as a cargo transport, were treated to a similar astonishing feat when a drone tracked cars over a highway near Quantico, Virginia.

One of two Model K-137s modified by Kaman during the early 1960s for drone evaluations. The unique helicopter is seen here at Baltimore's Friendship Airport during September 1964. (Frank Macsorely via Steve Miller)

Army and Marine enthusiasm over the prospect of drone cargo transport helicopters not only kept the program alive during the 1950s, it spurred the development of the "robocrane." The concept had an unmanned cargo helicopter ferrying supplies, over enemy territory if necessary, and returning to its ground operator, who used a suspended harness to direct the craft in proximity to the ground, much like leading a dog on a leash. During 1958 one of the HTK-1K drones was tested after its piston engine was replaced by a Task Corporation lightweight, high-voltage A.C. electric motor, powered by a cable plugged into a 250kw turbine-driven generator on the ground. The Army drone program eventually led to the use of drone aircraft as targets for Hawk missiles at White Sands Proving Ground, New Mexico.

Marine Corps interest in drones began in earnest during April 1954 when the Marine Corps Development Center formulated a study entitled, "Study of Marine Corps Requirements for Remotely Controlled Rotary Wing Aircraft." The study recommended that drones be acquired to evaluate their effectiveness as atomic weapons close support delivery systems, television camera-equipped intelligence gathering platforms, "umbrellas" for atomic aerial mines, battlefield illumination and cargo transport.

A similar proposal submitted by Kaman the following year envisaged broader applications to include minefield clearance, all-weather navigation and the all-important anti-submarine warfare.

During 1959 the Marine Corps Landing Force Development Center and HMX-1 evaluated the HTK-1K while determining Marine requirements for robot helicopters. Although the study was inconclusive, the Marine Corps, on 8 March 1960, altered policy to provide for the formation of one cargo helicopter drone squadron during 1963 and two more during 1964. The plan withered since drone helicopters proved to

Painted dark blue with orange trim, this QH-43G wore civil registration. Seen here on 27 May 1964, the drone is being pulled down to a ship's deck with a cable attached to the cargo hook. (Kaman)

be more expensive, less reliable and more difficult to operate than manned helicopters.

The advent of the turbine helicopter, combined with the Navy's keen interest in ASW, brought about the QH-43G. In a two-part highly classified project, the drone was used to test communications with submarines at sea, and proved that a homing weapon could be positioned directly over a sub, thus markedly increasing the kill probability. Serious consideration of a destroyer-based ASW capability began with preliminary trials aboard the USS WRIGHT and USS MITSCHER. On 23 May 1957 an HTK-1K made its first flight from the fantail of the Mitscher off Key West, Florida, demonstrating the feasibility of assigning torpedo carrying drone helicopters to destroyers.

In the sub communication mode, the QH-43G flew above the Wright, to which it was tethered by a long cable, while trailing a 10,000-foot, half-wavelength, low-amplitude antenna, which allowed radio communication with a submerged sub. Eventually, the system was replaced in lieu of a long antenna trailed from a fixed-wing aircraft.

Secret tests with the QH-43G extended well into the 1960s and ran concurrent with Kaman evaluations using two QH-43G equivalents, the Model K-137. Many of the principles pioneered by Kaman's drones were adopted for a modern fleet of Remotely Piloted Vehicles (RPVs), which enjoy a high rate of success.

**Under the direction of a shipboard operator, a modified HTK-1 drone is winched down to the deck of the destroyer USS MITSCHER in 1957 during anti- submarine warfare drone trials. (U.S. Navy)**

# Chapter 6

# The U.S. Air Force Huskie

Although the concept of air rescue dates back more than a century, it wasn't until World War II that it gained prominence. Recognizing the value of experienced pilots, Germany's Luftwaffe became the first to organize an air rescue unit, which began operations during the 1940 invasion of Norway. That same year, England's Royal Air Force and Royal Navy teamed up to organize their first air-sea rescue unit, which rescued aircrew downed during the Battle of Britain.

When the United States entered the war, it too established a rescue operation. However, rescue was given a low priority and was, therefore, largely unsuccessful. It wasn't until 1943 that an efficient rescue structure was in place, which, by war's end, had garnered save statistics worldwide. Though that success led to the birth of the Air Rescue Service (ARS) on 31 March 1946, under the Army Air Force Air Transport Command, air rescue was excluded as doctrine in military manuals. Its standing would remain subordinate after the U.S. Air Force was established during September 1947, especially in view of the government's forced austerity program, which continued through 1949.

The advent of the Korean war illustrated, once again, the need for air rescue capability. Using newer aircraft, specially trained rescue crews proved themselves an indispensable

**Kaman advertisement from 1963.**

**One of the small number of H-43As produced prepares for take-off at Craig AFB, Alabama during 1959. (Carl Damonte)**

**Manned by one pilot, a H-43A takes off from Craig AFB with a fire suppression kit. (Carl Damonte)**

The H-43A contingent at Craig AFB shares the flight line with T-33s. Both types were assigned to the 3615th Pilot Training Wing of the Air Training Command. The H-43As (serial numbers 58-1832 and 1835) served the unit from 1959 through 1962. (Carl Damonte)

part of tactical operations. As with World War II, the experiences gained in Korea were not applied directly to U.S. Air Force doctrine, and by 1957, air rescue was again fighting for its very existence. In an attempt to streamline its forces, the USAF reduced its Air Rescue Service from 12 groups and 39 squadrons to 7 groups and 17 squadrons.

The limitations and expense of dispersing worldwide rescue services among numerous agencies forced a decision based on what was already learned after World War II; that a highly proficient rescue force under one command would result in a more efficient operation. As the Air Rescue Service maneuvered to modernize and expand its operation during the early 1960s, it was pressed to provide aircraft to support

the National Aeronautics and Space Administration (NASA), and shortly after, support for tactical air operations in Southeast Asia's rapidly expanding war.

Numerous advances in rescue techniques and operations have been recorded during the history of the Air Rescue Service. At the forefront of many of those accomplishments was Kaman's Huskie. However, its introduction to the Air Force paralleled the roller coaster beginnings of the commands to which most were assigned.

During a period when air rescue appeared to be on the upswing, the U.S. Air Force recognized the need for a helicopter to perform its newly defined "Local Base Rescue" mission. Paramount among the requirements was the helicopter's

Prior to the H-43's arrival, Air Force helicopter rescue operations were handled primarily by two types, one of which was the Piasecki H-21, pictured here. (Author's collection)

Besides the H-21, Sikorsky H-19s served in the rescue role until the Huskie's arrival. This example was designated a SH-19B. (Dale Mutza)

The first H-43B to become operational with the Tactical Air Command was number 58-1858, seen here at Luke AFB, Arizona in April 1960. A TAC insignia is displayed on the tail fin which is part of the early tail configuration. (Kaman)

The color scheme of painted aluminum with International Orange trim became the standard for Air Force Huskies until 1963. (Kaman)

ability to maintain an "alert" posture, to fly to a crash site with personnel and equipment to suppress fire and save lives. Kaman entered its proven HOK-1/HUK-1 design in the 1956 competition.

Souring the competitive flavor of the event was a controversy that went beyond the limits of inter-business rivalry and ethics. Two Air Force majors, who made known their preference for an aircraft that was easy to fly, stable and hovered well (attributes of Kaman's design), caused others in the competition to incorrectly assume the pair had been bribed. That led to allegations to a commanding general, who summoned the FBI to investigate. Since the assertions threatened Kaman's integrity, Charles Kaman himself met with two Air Force generals at the pentagon.

Only Huskies that operated near large bodies of water were equipped with emergency float gear. This HH-43B belonged to the 40th ARRWg. (MAP)

This HH-43B wears the ATC emblem on the tail. (Skip Robinson)

Visible in the cabin of this H-43B is a "Pugh" rescue basket, which had to be trimmed to fit through the cabin door. (Jack M. Friell)

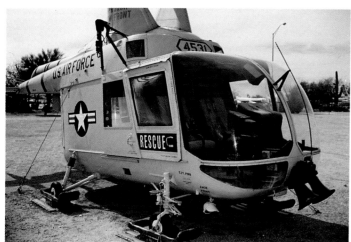

A HH-43B finished in overall light gray. (Dale Mutza)

Only a handful of Huskies served the Air Material Command. This H-43B was assigned to the logistics depot at Griffiss AFB, Rome, New York. (MAP)

The high-ranking pair admitted having heard of control problems with Kaman's helicopter, prompting Kaman to ask that they each fly in a HOK-1 with a Marine pilot. Kaman was certain that would disprove the allegations, explain the enthusiasm of the two majors and gain the generals' confidence, as fixed-wing aviators, in Kaman's product. They agreed, and the Navy made arrangements for the flights at MCAS New River. Each general was flown simultaneously by a Marine lieutenant, who passed the controls to his general within the first few minutes of flight. After an hour, the generals returned with big smiles, the investigation quickly ended and the competition entered its final stages.

Kaman won with its HOK-1/HUK-1 configuration, and was awarded a contract for the production of 18 H-43As. After the aircraft were flight tested, they were assigned to Air Training Command bases. During December 1956 Kaman submitted a proposal to the Air Force for 48 piston-powered H-43As and 19 turbine-powered H-43Bs. However, when tests begun three months earlier proved the feasibility of converting the H-43 to turbine power, the switch was made to an all-turbine line.

When H-43s began to appear during 1959, flyers at more and more bases took comfort in the fact that quick rescue had become a reality. Prior to the H-43's arrival, Sikorsky SH-

During 1963, this HH-43B was used to drop a dummy warhead from 10,000 feet into the bay near Seattle, Washington. The Huskie was controlled by radar to the exact release point. The Department of Defense test, on behalf of the Navy, was conducted to determine penetration characteristics of a warhead on a hard target. (William A. Luther)

Thanks to pilot proficiency and the Huskie's autorotation capabilities, this aircraft was saved from destruction when it suffered a catastrophic engine failure at low altitude and low airspeed. Captain John Christianson of Det. 5, 40th ARRWg earned the MAC Outstanding Individual Safety Award for his action during the forced landing. He went on to pass the 2,000-hour mark in the same aircraft. (John Christianson)

Framed by Douglas DC-3s, this HH-43B, wearing the MATS emblem, was displayed at Le Bourget, Paris during June 1965. (Jack M. Friell)

19A/Bs and Piasecki SH-21s gave the Air Force a limited rescue helicopter capability. The first rescue with an Air Force H-43B occurred on 24 April 1960 at Superstition Mountain, Arizona. Although several blade-to-tail strikes grounded the H-43Bs during the early 1960s, one aircraft was cleared to take on a rescue of mountain climbers in Alaska's forbidding terrain. The successful effort foreshadowed the H-43's impressive Air Force career.

Most of the initial production batch of 98 H-43Bs was evenly distributed among the Strategic Air Command and Air Defense Command, 34 and 33 aircraft respectively. The remainder were dispersed among the following: Air Research and Development Command, which received 6; the Air Material Command took delivery of 10; the Military Air Transport Command received 5; and 6 went to the Air Training Com-

mand. Four B models joined all H-43As in the Tactical Air Command. Since the H-43 displayed excellent performance characteristics and versatility, early assignments under those commands often contrasted sharply with rescue, the mission for which it was originally intended. During March 1960 the first H-43B earmarked for tactical use was ferried from the factory to Francis E. Warren AFB, Wyoming. There, H-43s conducted missile site evaluations for the Strategic Air Command until June 1961, when the type was transferred to the Air Rescue Service under the Military Air Transport Command (MATS). During March 1961 the Air Force ordered three H-43Bs for export to satsify the first of many Military Assistance Program requirements.

During 1961-62, when sufficient numbers of H-43Bs were on hand, the H-43As were placed in "flyable storage" at the

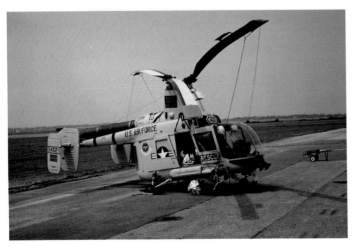

Unusual on this Europe-based HH-43B is the rotor warning on a black background. (Jack M. Friell)

The flight mechanic (crew chief) is perched in the cabin doorway of this Huskie based in England. (Jack M. Friell)

This Huskie is seen in terrain that precluded operations by most aircraft types. The HH-43B is fully in its element in Greenland during January 1968. (Don Spering/A.I.R. Collection)

Military Aircraft Storage and Disposition Center, Davis-Monthan AFB, Arizona. Under the revised designation system of 1962, the H-43B became the HH-43B, the additional "H" prefix signifying Search and Rescue (SAR). During the same period, airmen nicknamed it the "Huskie"—Kaman steadfastly used the spelling "Husky." Like its canine namesake, the sturdy helicopter proved its adaptability to multiple tasks in extreme climates, terrains and altitude.

Not only was the Huskie the USAF's first turbine-powered helicopter and the first built specifically for the Local Base Rescue role, it quickly distinguished itself in the record-setting arena. In the hands of Air Force pilots, the Huskie achieved numerous performance and endurance records. To demonstrate its climbing ability, an H-43B was flown to a world's helicopter record of 30,000 feet in 1959 by Majors William J. Davis, Jr., and Walter F. Hodgson. Davis worked with the Huskie at the Aeronautical Systems Center and Hodgson was the Air Force H-43 Project Officer at Edwards

AFB Flight Test Center. The H-43B serial number 58-0263 established seven world records during 1961-62 for helicopters in its class for rate of climb, altitude and distance traveled. The Huskie was assigned to rescue duty with Detachment 3, 42nd ARRS, Kirtland AFB, New Mexico, prior to its retirement and flight to the Air Force museum, Wright-Patterson AFB, Ohio, in April 1973.

Shortly after it entered Air Force service, the Huskie participated in programs that ranged from mundane to the unusual. Beginning in 1958, the Air Rescue Service took on the added responsibility of mid-air recovery of space hardware and other items that descended from parachutes. Extensive trials with fixed-wing aircraft proved that helicopters would be better suited for the aerial recovery of large objects at slower speeds. Retrieval by helicopter also spared sensitive components from the harsh forces common to fixed-wing recovery. After several private tests, the first funded system using a helicopter became operational at NASA Wallops Island during late 1960. Its success prompted All American Engineering (AAE), who designed the system, to convince the Air Force that they needed helicopter recovery capability.

A contract was awarded AAE during 1961 for a system capable of recovering 800 pounds with a H-43B helicopter. The aircraft's rear doors were removed and a rig was mounted to the aft cabin floor. A round hole was cut in each fuselage side through which 25-foot long poles extended rearward at an outward angle. The poles were raised and lowered with a hydraulic hand pump. A Model 80C winch was mounted on a pallet with its cable leading forward to a pulley on the cargo hook, then aft over rollers. The cable was then fastened to a nylon loop which incorporated a dangling hook.

Since balloon and drone operations were conducted at Holloman AFB, New Mexico, the system was delivered there during July 1961. During the first aerial test, the operator in the H-43's cabin snagged a descending parachute on the first pass. During additional tests, flown between 55 and 65 mph, 13 objects, weighing between 100 and 800 pounds, were snatched in mid-air.

Finished in the final non-combat scheme, this HH-43B displays part of its rescue gear during 1969. (Nick Williams)

HH-43B number 62-4538 and its fire suppression kit at Le Bourget during 1967. (Nick Williams)

A HH-43B during May 1964. Markings include the MATS emblem and an Outstanding Unit Award above the copilot's door. (Dave Menard)

This HH-43B on ready-alert at Webb AFB, Texas during 1967 has a sign in the cockpit which reads, "Entry Prohibited Helicopter Cocked." Firefighters' gear is poised outside the cabin. (Steve Miller)

A great deal was learned about parachute design requirements for mid-air recovery during the H-43 trials, which led to a successful aerial recovery program. The Air Rescue Service was tasked with the search for and recovery of space hardware and support of all manned space missions beginning in 1961. In view of its expanding role in space flight support, the Air Rescue Service was re-designated the Aerospace Rescue and Recovery Service (ARRS) on 8 January 1966.

During the latter part of that decade, more than 100 ARRS units operated from nearly 90 locations in the U.S., Guam, the Canal Zone, and 14 foreign nations. Besides 18 squadrons and approximately 70 detachments in the U.S. and overseas, the ARRS also operated several joint search and rescue centers for the unified commands. Search and rescue facilities, which were strategically located to meet civil and military rescue needs, were governed by ARRS headquarters at Scott AFB, Illinois, through three rescue wings: the 39th ARRWg at Eglin AFB, Florida; the 40th ARRWg at Ramstein AB, Germany; and the 41st ARRWg at Hickam AFB, Hawaii.

Continental search, rescue and recovery operations were coordinated by the 39th ARRWg through one of three SAR centers: the 44th ARRSq/Eastern Aerospace Rescue and Recovery Center (EARRC) at Eglin AFB; the 43rd ARRSq/Central Aerospace Rescue and Recovery Center (CARRC) at Richards-Gebaur AFB, Missouri; and the 42nd ARRSq/Western Aerospace Rescue and Recovery Center (WARRC) at Hamilton AFB, California.

Detachment 1 of the 39th ARRWg at Elmendorf AFB, Alaska, was the joint SAR center for the Alaskan Air Command, and Detachment 2, at Albrook AFB, Canal Zone, served the U.S. Air Force's Southern Command. Search and Res-

This view of serial number 60-288 reveals the HH-43B's underside details. (Steve Miller)

Parked next to number 60-288 is a Bell twin-engine "Huey", two versions of which replaced the Air Force Huskie. (Esposito via Dave Menard)

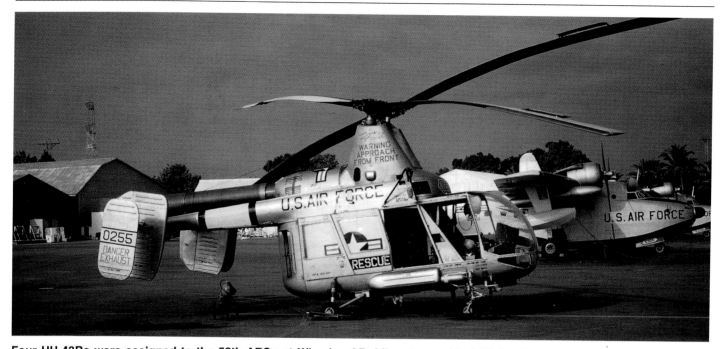

Four HH-43Bs were assigned to the 58th ARSq at Wheelus AB, Libya to provide LBR for Europe-based aircraft that rotated through Wheelus and El Votia Bombing Range. They were also involved with SAR in the Mediterranean and supported the Water Survival School. U.S. operations at the base ceased by June 1970 when control of the sprawling facility was returned to the Libyan government. Both Arabic and American warnings were applied to the rotor pylons. (Jack M. Friell)

cue activities in Europe were handled by the 40th ARRWg/Atlantic Aerospace Rescue and Recovery Center (AARRC), and in the Pacific by the 41st ARRWg/Pacific Aerospace Rescue and Recovery Center (PARRC). The 3rd Aerospace Rescue and Recovery Group (which came under control of the 41st ARRWg) at Tan Son Nhut AB, Vietnam, became the hub for all SAR operations in Southeast Asia.

It wasn't long before the highly adaptable Huskie reached milestones in Air Force service. Serial number 59-1543, one of two H-43Bs attached to Det. 41, EARRC, Loring AFB,

Maine, became the first ARS Huskie to reach the 1,000-flying hour mark. On 17 November 1966, HH-43B s/n 59-1557 logged its 2,000th flight hour on a search mission for a lost hunter, making it the ARRS's high-time Huskie. Credit for the 2,000th LBR scramble went to HH-43B s/n 60-261 of Det. 15 WARRC at Luke AFB, Arizona, on 29 December 1964. That year saw an all-time high in rescue work as the Huskie detachment made 13 pilot bailout recoveries and flew 51 rescue missions involving 99 civilians. The detachment was so heavily relied upon that refueling points were set up throughout Arizona, allowing missions beyond its borders.

Huskie alert pads included yellow and red concentric circles on the ground and precise placement of the aircraft, fire suppression kit and positioner pole. Pictured here is the pad at Hahn AB, Germany. (John Christianson)

Jet jockeys didn't get all the glamorous assignments. These ROTC "Angel Flight" students are obviously captivated by Captain John Christianson's tales of Huskie adventures. (John Christianson)

**Personnel and Huskie of Det. 12, CARRC at Randolph AFB, Texas during 1970. (U.S. Air Force)**

**The H-43B number 58-1857 crashed at Warren AFB, Wyoming during July 1960. A bolt installed upside down brought the aircraft down. (U.S. Air Force)**

Similar refueling sites were established for Huskies serving abroad. To further extend the outstretched arms of 31st ARRSq HH-43Bs at Clark AB, Philippine Islands, helipads and fuel storage areas were set up at Cubi Point NAS and Sangley Point NAS. An increase in the number and range of missions by 1966 necessitated the establishment of similar sites at John Hay AB, Wallace AS, Paredes AS and a Philippine Air Force radar site.

The Huskie's diversity was illustrated by a pair of HH-43Bs (s/n 58-1844 and 59-1548) assigned to Det. 1, Air Force Special Weapons Command at Indian Springs AFAF, Nevada. Their primary mission, which was to support the Nevada Test Site and Nuclear Rocket Development Station, included photography, standby search and rescue and VIP flights. The special detachment also flew missions which were among the highest ever flown by Huskies. Six consecutive flights were made above 20,000 feet, the most demanding of which involved lifting an 800-pound external payload to 22,300 feet.

Flights to 16,000 feet at the facility were common. Though not a rescue unit, the detachment provided support for various emergencies, along with rescue and investigation activities at Nellis AFB.

Huskie crews added to their list of persons saved when they joined forces with the U.S. Army during 1972 in a joint Department of Defense-Department of Transportation pro-

**Serial number 59-1567 was only 28 days old with 104 hours when it crashed during October 1960 due to a control reversal situation. (Dick Van Allen)**

**Huskie crewmen's view of a person being brought up on the rescue hoist. (John Christianson)**

The U.S. Air Force's entry into the 6th International Rescue Helicopter Competition at Woodbridge AB, England during 1972, HH-43B number 62-4521, takes its place on the flight line among other competitors. The Royal Air Force edged the USAF team from the 40th ARRWg, which placed second overall, but won the scramble event. (John Christianson)

gram that used military helicopters to evacuate victims of auto accidents. Initially, two Huskie detachments participated in the program, which was called Military Assistance to Safety and Traffic (MAST). Before long, Huskie units across the nation were flying MAST missions, often under adverse conditions, which ranged from saving the lives of premature babies in small mining towns to flying heart attack victims out of remote areas. The MAST program provided the impetus for today's widespread use of helicopters for Emergency Medical Services (EMS).

Rescue became synonymous with the Huskie as it amassed large numbers of lives saved. During 1967 alone, 52 percent of total combat zone rescues in Southeast Asia were made by HH-43s, which also accounted for 54 percent of worldwide ARRS rescues. Those statistics were not without losses, which underscored the hazards of a profession where the rescuer sometimes became the victim.

No period in the Huskie's history reflected that more than 1969, a dark year for the Military Airlift Command, which recorded 13 major and 2 minor aircraft accidents. The Huskie,

A Huskie displays its wide variety of equipment. The large MA-1 rescue basket was used for pickups from land or water. (Bob Collom)

Huskie crewmen display the FSK and "horse collar" and forest penetrator rescue hoist attachments. The HH-43B, serial number 58-1854 wears only its last three s/n digits on the fuselage. This was the alert pad of Det. 12, 42nd ARRSq at George AFB, California during 1971. (Jim Burnett via Robert F. Dorr)

A HH-43B of the USAF Helicopter School at Sheppard AFB hovers over firefighter trainees as they prepare to attack a large spill fire. (U.S. Air Force)

otherwise known for its admirable safety record, was largely responsible for the gloomy statistic. First to fall victim was a HH-43B which crashed at a U.S. base shortly after it was scrambled to an emergency. The suspended fire suppression kit forced the Huskie to fly at an unsafe altitude when the engine failed. A mere five feet was considered the safe altitude for single-engine helicopters operating near the ground.

The next incident involved a HH-43B returning to base when lightning struck the rotor blades, causing them to disintegrate. All eight aboard lost their lives when the Huskie exploded and burned on impact. Fate loomed over another Huskie crew that investigated dense smoke near a Far East base; the aircraft's low approach at low speed in a downwind condition caused it to hit the ground and roll over.

A HH-43B of Det. 8, 44th ARRSq, attached to the 354th Tactical Fighter Wing at Myrtle Beach AFB, practices medical evacuation during 1971. (via Robert F. Dorr)

A Huskie from the original production batch of H-43Bs assigned to the Air Training Command. (U.S. Air Force)

At Utapao Royal Thai Air Force Base, a HH-43B was scrambled to a bomb-laden B-52 bomber that had aborted a heavy-weight takeoff and ran off the end of the runway. With its forward landing gear collapsed, the crippled bomber started to burn. The Huskie hovered briefly near the bomber's nose, then moved off to hold three to four hundred yards from the scene. Unable to confirm that the B-52's tail gunner was clear, the HH-43 moved back in for a fly-by when the bomber exploded. The entire rotor system was ripped from the Huskie, which crashed, killing three of the crewmen. It was later discovered that the tail gunner had escaped.

During high altitude operations at the mountain crash site of a commercial airliner, a Huskie became enveloped in a downdraft during a shallow approach. With recovery from the sink-rate impossible, even with maximum power, the rotor blades struck rocks and the Huskie tipped over.

Engine failure in a single-engine helicopter over land is bad enough, however, fling-wing flyers will quickly agree that a flame-out over water greatly increases "pucker factor." One crew endured the unpleasant experience while delivering a HH-43B from one detachment to another in Southeast Asia. Tactical operations forced the coastal flight a few miles off-

A HH-43B makes a water pickup during early 1966. (U.S. Air Force)

Huskies routinely carried out medical evacuation missions to naval ships. Pictured here is serial number 58-1843, one of the first H-43Bs produced. (U.S. Air Force)

shore when the engine quit. Unable to restart during autorotation, the chopper was ditched in the ocean. The crew escaped and was rescued, but the sea claimed the Huskie.

As if combat operations in Southeast Asia weren't enough to compile a substantial number of HH-43 losses during 1969, "operator factor" added its share. After completing a base perimeter check, a HH-43B crew practiced autorotations with the final maneuver near a BAK-12 barrier. After landing, the pilot rolled over the barrier cable, which snagged the Huskie's landing gear. The aircraft clumsily rolled onto its side, causing the rotor blades to strike the ground and disintegrate.

The final non-combat Huskie loss of that fateful year occurred after a HH-43B took off from a jungle clearing. Upon liftoff, the engine suddenly lost power, and the pilot autorotated into the trees. As testimony to the Huskie's renewed reputation as the safest helicopter to fly, serial number 58-1852 became the first HH-43 to log 5,000 accident-free hours on 12 May 1972, while assigned to Det. 5, 42nd ARRSq.

Training in the H-43 was conducted at the unique U.S. Air Force helicopter school, which had its meager beginning during January 1944 at Freeman Field, Seymour, Indiana. After a number of setbacks, the school earned the title "Gypsy School" following a series of migrations—during its first quarter century of service, the school had moved nine times.

Since Army pilots comprised nearly 90 percent of the students, the Air Force suffered a critical shortage of instructor and operational pilots. When the Army established its own program during July 1956, the Air Force school moved to Randolph AFB, Texas, where the training syllabus centered around the H-13, H-19 and H-21 helicopters. An advanced mountain training program was set up at Stead AFB, Nevada, where terrain and climate proved so valuable that the entire

school was moved there in 1958. The HH-43 was added in April 1960 to become the school's first turbine aircraft.

Would-be pilots first received training in the North American T-28 "Trojan" at Randolph, then went on to Stead for rotary-wing school. Helicopter flight training was broken down into 70 hours in the UH-19B, and 35 hours in either the H-43B or newly arrived CH-3. The Huskie course included a five-hour phase devoted to fire suppression. Students from nations that operated Huskies also attended the school under the Military Assistance Program.

In addition to training students, Stead's helicopters played a vital role in support and rescue missions that ranged from

Having dropped off its fire suppression kit and firefighters, a Huskie positions to beat back flames as firefighters move in. (U.S. Air Force)

In a scene repeated countless times by Huskie crews worldwide, a ready-alert HH-43B is "scrambled" for an emergency. (U.S. Air Force)

A Huskie takes part in medevac training for aeromedical technicians at Brooks AFB, Texas during February 1969. (National Archives)

survey flights of national parks with the Secretary of the Interior, to the rescue of three seriously injured persons whose light plane crashed in the High Sierra. From December 1964 to January 1965, 27 of the school's helicopters participated in the rescue of thousands during northern California flooding.

When the school was transferred from Stead to Sheppard AFB, Texas, during 1966, a new concept had undergraduate pilots earning their helicopter wings, having been awarded their fixed-wing rating through a separate program.

Helicopter pilot training often proved to be a traumatic experience for unsuspecting fixed-wing pilots. After discovering that a helicopter's many controls warranted constant attention and coordination, the student overcame his fear that his aircraft would fall out of the sky. With time and effort, the student learned to trust the helicopter and actually enjoy the experience. As he neared status as a complete convert, he even adopted the school motto, "Help stamp out fixed-wing."

With the introduction of the Bell UH-1F "Huey" to the school during May 1967, student pilots first trained in the type before transitioning to the HH-43B. Throughout Huskie training, firefighter training was given in conjunction with pilot training to develop, from the beginning, the teamwork necessary between crewmembers. Also vital to the Huskie team were the mechanics, who learned their trade through an initial 14-week basic helicopter school, followed by an intense 6-week course on the Huskie.

By the end of the decade, specialized ARRS aircrew training was carried out at Hill AFB, Utah, by the 1550th Aircrew Training and Test Wing (ATTW), which included the consolidated helicopter school.

By 1970 plans were made to replace the venerable Huskie with a more effective and versatile rescue helicopter. The obvious replacement was Bell's HH-1H Huey, whose twin-engine cousin, the UH-1N, had already joined the Air Force as part of a helicopter modernization program. The first of 30 HH-1Hs was delivered to the 1550th ATTW during early February 1971, followed in mid summer by deliveries to rescue units at Holloman, K.I. Sawyer, Mountain Home, Plattsburgh and Edwards Air Force Bases. Helicopter pilots who transitioned into the higher-powered Huey found its flight characteristics considerably different from the Huskie in several crucial areas. For all the improvements the Huey did offer, it couldn't match the Huskie's hover performance at high altitudes and heavy gross weights, nor were its autorotation characteristics as favorable. Air Force chopper pilots who strapped into a sleek, powerful Huey never lost their respect and admiration for the able Huskie.

From Bell Helicopter's line of "Hueys" came the HH-1H, which replaced the Huskie beginning in 1971. (Candid Aero Files)

# Chapter 7
# Crash-Rescue

On 24 September 1961—When Captain Jack C. Armstrong flew his H-43B Huskie from Seymour-Johnson AFB to the municipal airport at Wilmington, North Carolina, for the annual air show, little did he realize that he and his crew would play a major part in an unscheduled drama before thousands of spectators. The Huskie was placed on display, the crew explained their crash-rescue duties to the curious, and sat back to take in some of the air show.

In preparation for a mass jump, the U.S. Army parachute team boarded the C-123 transport flown by the "Thunderbirds" flight demonstration team. The Huskie crew watched the C-123 lift off the runway, climbing steeply—so steeply that the H-43 crew instinctively sensed trouble and dashed for their

Huskie. As Captain Armstrong reached the cockpit, he heard the sickening crunch of metal impacting concrete and turned to see flames erupt around the crashed transport. A minute and a half later, the Huskie was speeding the few hundred yards to the crash scene.

Coming to a hover in front of the burning C-123, Armstrong observed that none of the 15 occupants had escaped. Just as he began to land to drop off his firefighters, fire enveloped the transport's cockpit. He repositioned the Huskie, his rotor downwash suppressing the flames while supplying air to trapped victims and rescuers working in the burning fuselage to extricate them.

After maintaining a hover for 20 minutes, Armstrong touched down to drop off the two firefighters and flight sur-

**The third HOK-1 built carries both Air Force and Marine markings during Air Force trials during May 1958. Suspended from the cargo hook beam is the original fire suppression kit. (Kaman)**

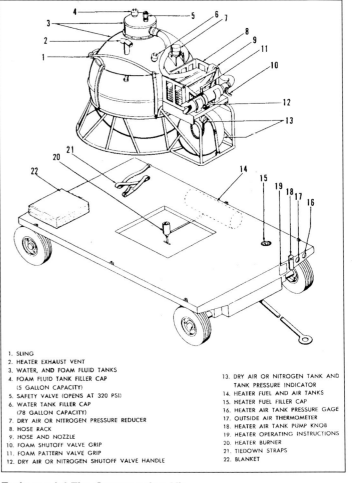

1. SLING
2. HEATER EXHAUST VENT
3. WATER, AND FOAM FLUID TANKS
4. FOAM FLUID TANK FILLER CAP
   (5 GALLON CAPACITY)
5. SAFETY VALVE (OPENS AT 320 PSI)
6. WATER TANK FILLER CAP
   (78 GALLON CAPACITY)
7. DRY AIR OR NITROGEN PRESSURE REDUCER
8. HOSE RACK
9. HOSE AND NOZZLE
10. FOAM SHUTOFF VALVE GRIP
11. FOAM PATTERN VALVE GRIP
12. DRY AIR OR NITROGEN SHUTOFF VALVE HANDLE
13. DRY AIR OR NITROGEN TANK AND
    TANK PRESSURE INDICATOR
14. HEATER FUEL AND AIR TANKS
15. HEATER FUEL FILLER CAP
16. HEATER AIR TANK PRESSURE GAGE
17. OUTSIDE AIR THERMOMETER
18. HEATER AIR TANK PUMP KNOB
19. HEATER OPERATING INSTRUCTIONS
20. HEATER BURNER
21. TIEDOWN STRAPS
22. BLANKET

**Early model Fire Suppression Kit.**

**A H-43A hovers protectively over a team of firefighters as they enter a fuel fire. (U.S. Air Force)**

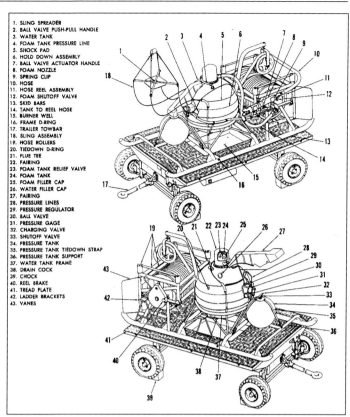

1. SLING SPREADER
2. BALL VALVE PUSH-PULL HANDLE
3. WATER TANK
4. FOAM TANK PRESSURE LINE
5. SHOCK PAD
6. HOLD DOWN ASSEMBLY
7. BALL VALVE ACTUATOR HANDLE
8. FOAM NOZZLE
9. SPRING CLIP
10. HOSE
11. HOSE REEL ASSEMBLY
12. FOAM SHUTOFF VALVE
13. SKID BARS
14. TANK TO REEL HOSE
15. BURNER WELL
16. FRAME D-RING
17. TRAILER TOWBAR
18. SLING ASSEMBLY
19. HOSE ROLLERS
20. TIEDOWN D-RING
21. FLUE TEE
22. FAIRING
23. FOAM TANK RELIEF VALVE
24. FOAM TANK
25. FOAM FILLER CAP
26. WATER FILLER CAP
27. FAIRING
28. PRESSURE LINES
29. PRESSURE REGULATOR
30. BALL VALVE
31. PRESSURE GAGE
32. CHARGING VALVE
33. SHUTOFF VALVE
34. PRESSURE TANK
35. PRESSURE TANK TIEDOWN STRAP
36. PRESSURE TANK SUPPORT
37. WATER TANK FRAME
38. DRAIN COCK
39. CHOCK
40. REEL BRAKE
41. TREAD PLATE
42. LADDER BRACKETS
43. VANES

**Late model Fire Suppression Kit.**

geon. He again took up a hover position, then landed when 12 survivors were removed and rushed two of them to a nearby hospital. One of the survivors aboard was given a life-saving tracheotomy enroute to Wilmington by a doctor who had jumped aboard. The Huskie made a vertical descent on a tennis court beside the hospital, its rotors clearing the building by only a few feet.

15 December 1960—A B-52 bomber experienced an inflight refueling mishap and headed for Larson AFB. Meanwhile, a base H-43B crew was scrambled and held a standoff position near the runway. As the massive, crippled bomber lumbered down the runway, the right wing was torn off and a huge fuel fire erupted. As explosions rocked the scene, the Huskie closed in and hovered over the cockpit section of the

**The first-built H-43A lifts its FSK from a trailer. On the succeeding "B" Model, the rescue hoist was relocated to the aircraft's right side and the cargo hook beam was replaced by a hook recessed into the belly. (Kaman)**

burning fuselage, driving back the smoke and flames, allowing ten crewmen to escape. At one point, an explosion blew the Huskie off position, however, it recovered and was repositioned to continue beating flames and smoke away from the bomber's nose.

7 July 1961—An F-89 Scorpion jet radioed a "Mayday" with an inflight engine fire. On approach to Connally AFB, the trailing fire was observed to be four times larger than the aircraft itself. A H-43B followed the flaming F-89 down the runway to a stop and immediately positioned so its rotor downwash blew the fire away from the cockpit, long enough for the Scorpion's pilot and radar observer to escape.

2 November 1960—An engine of a KC-97 caught fire as the Stratotanker was inbound to Randolph AFB. A H-43A was scrambled and followed the KC-97 to a crash landing in a field beyond the runway. The H-43A crew extinguished the engine fire before 4,000 gallons of fuel aboard the tanker ignited. The rescuers then assisted 11 crewmen from the crash and flew them to the hospital. It wasn't until the H-43A's third trip to the site that ground vehicles were able to arrive.

Such scenes were repeated so often throughout the H-43's life that their documentation would fill volumes.

It all began with a government analysis of military aircraft accidents for the two-year period 1955 to 1957. The study

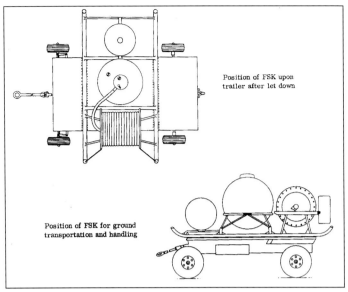

Position of FSK upon trailer after let down

Position of FSK for ground transportation and handling

**FSK positions for ground and air handling.**

**A H-43A backs up firefighters during a training exercise at Craig AFB, Alabama during 1959. (Carl Damonte)**

showed that of 986 crashes, 952 (or 96.6 percent) occurred within a 61-mile radius of the air base. Most of those (72 percent) actually occurred very close to the base—within ten miles—and 46 percent resulted in a crash fire. There was an obvious need for a fast and reliable method of rescuing aircrew from crashes that occurred in places inaccessible to standard ground fire apparatus. The study went on to outline the requirements for a small, high performance helicopter that could respond quickly, perform hoist rescue, carry firefighting and rescue personnel, and serve as an aerial ambulance.

Preliminary tests with the Marine HOK-1 on small, 200-gallon fuel spill fires had taken place during 1956. At the conclusion of those tests, and recognizing the success of the U.S. Navy's use of helicopters for plane guard duty during aircraft carrier flight operations, the USAF formulated its Local Base Rescue (LBR) mission. When they chose Kaman's

H-43 to carry out that mission, the Air Force soon discovered that they got more than they bargained for.

During initial tests with fuel fires, a phenomenon came to light, which became an added benefit in the H-43's firefighting role. Kaman's counter-rotating rotors displayed a unique ability to produce huge volumes of low velocity air with a longitudinal forward thrust, which opened a "cool" corridor through a fire. Firefighters had actually walked through the corridor with minimal protective clothing and received no burns. The rotor-produced corridor allowed pilots to form a path to an aircraft fuselage, through which firefighters could fight the fire from inside out, thereby increasing the possibility of rescuing trapped aircrew. Since the mission was to save lives, it was more important to clear a path and control the fire, rather than attempt to extinguish the entire blaze. The front-directional rotor wash also cooled the firefighters below and helped the spread of extinguishing agents.

**How the Huskie's fire suppression kit got its nickname "Sputnik" is obvious in this view of the unit perched upon its trailer.**

**With the FSK deployed and firefighters advancing toward an aircraft fuselage, a H-43B uses its rotor downwash to provide cool air, push away smoke and flames and help spread the foam blanket. (Kaman)**

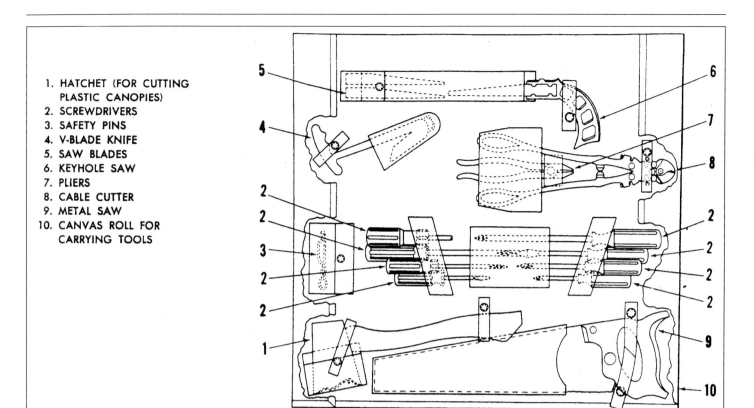

1. HATCHET (FOR CUTTING PLASTIC CANOPIES)
2. SCREWDRIVERS
3. SAFETY PINS
4. V-BLADE KNIFE
5. SAW BLADES
6. KEYHOLE SAW
7. PLIERS
8. CABLE CUTTER
9. METAL SAW
10. CANVAS ROLL FOR CARRYING TOOLS

Crash entry kit.

Its FSK attached, a HH-43B clears the alert pad on a scramble mission. Dual loudspeakers and searchlights were standard equipment on LBR Huskies. (U.S. Air Force)

As the Huskie nears the positioner pole, the FSK is attached to the cargo hook. The ground crewman will then knock down the pole and signal the pilot up. (Kaman)

A H-43B clears a rescue path for firefighters. Later in the Huskie's career, the rear clam-shell doors were removed and a nylon cargo net used to facilitate movement through the large opening. (Kaman)

On 24 September 1961, a C-123 transport crashed on takeoff during an airshow at Wilmington, North Carolina. A H-43B (s/n 60-269) was one of the aircraft on display for the event. It flew to the crash site and was credited with saving 12 survivors. (Kaman)

In conjunction with its goal to consolidate all rescue functions under one command, the Air Force transferred the Local Base Rescue mission from the jurisdiction of various commands to the Air Rescue Service during December 1960. On 1 October 1961, the ARS organizational chart included 70 LBR units. By year's end, a LBR unit was assigned to every major Air Force installation. The ARS concept called for each LBR unit to consist of two aircraft, four officers and seven airmen, with one helicopter and crew on 24-hour ready-alert. The H-43B served as the primary LBR helicopter, 69 of which were operational by the end of 1961. Besides 17 H-43As, the unique helicopter force also included 58 H-19Bs and 4 SH-

21Bs. Neither was equipped for fire suppression and were phased out with the arrival of new Huskies. As more H-43s became available, LBR units at large bases, or those with heavy traffic, were assigned three aircraft.

Initially, a H-43 crew comprised a pilot, firefighter and crash-rescue medical technician. Training and experience quickly brought about an increase in the crew, which comprised a pilot, copilot, flight engineer, medical technician and two firefighters. All crewmen became proficient in land and water rescue since LBR aircraft were often sent to non-aircraft emergencies off base. For the rescuemen, the rewards far outweighed the long hours worked.

The Huskie pilot's view of firefighters working their way through a training fire at Nha Trang AB, South Vietnam during 1968. (John Christianson)

This view clearly illustrates the unique ability of the Huskie's rotor system to push smoke and flames away from a burning aircraft. (John Christianson)

Cap and unofficial wings worn by HH-43 firefighters. (Author's collection)

A Huskie prepares to go to work on a practice spill fire.

In an aircraft accident, with highly flammable fuels and often armament on board, there was no time to waste. A typical crash-rescue scenario had the control tower alerting the LBR detachment that an aircraft in trouble was inbound. The Huskie, its turbine engine requiring no warmup, was airborne in less than 90 seconds. It hovered over a nearby Fire Suppression Kit (FSK), which was attached to the helicopter's external cargo hook. Orbiting the base, the Huskie met the aircraft in trouble and followed it during its landing run. In the event of a crash or fire, the FSK was placed on the ground and automatically detached from the cargo hook. After offloading the firefighters, the Huskie began a hover ten feet

Having just picked up its FSK, a HH-43B heads out to intercept an aircraft that radioed an emergency in Southeast Asia. (U.S. Air Force)

off the ground with its rotor system at the fire's edge. Firefighters deployed the FSK hose, activated the pressure system, and began laying a foam path to the burning aircraft. Expert use of the Huskie's controls enabled a pilot to use the rotor wash to blow/roll the foam path along. Once they arrived at the aircraft fuselage, firefighters used their skills in locating crash exits, disarming ejection seats, removing injured persons and cutting off power supplies.

If no fire developed, the FSK was left at the scene while injured persons were airlifted to a hospital. Enroute, they were treated by the onboard medical technician. Not only were Huskie medics qualified in emergency medical care, they also had received additional training in helicopter rescue techniques.

Since no two crashes or fires are alike, crash-rescue crews trained endlessly to meet and adapt to any emergency. Firefighters learned to work beneath the helicopter on different size fires on various terrain surfaces, while pilots honed their hover skills to a sharp edge. Many incidents occurred at night or in bad weather, while other predicaments, especially those involving large aircraft, required two Huskies. Through constant practice, flight crews and firefighters maintained the coordination necessary to attack a fire to rescue airmen who would otherwise be trapped in burning wreckage. The Huskie and its crew were so heavily relied upon that a HH-43F (s/n 60-289) was assigned to Maryland's Andrews AFB, where it flew disaster cover at every takeoff and landing of "Air Force One."

Kaman initially supplied a rudimentary fire extinguishing unit for its line of helicopters. However, the H-43's FSK was designed and tested at the Wright Air Development Division,

In its element - a H-43B, with an early tail configuration, moves to the upwind side of a blaze. (Kaman)

Images such as this leave no doubt as to why "Firebird" became a common call sign for Huskies stationed in the U.S.

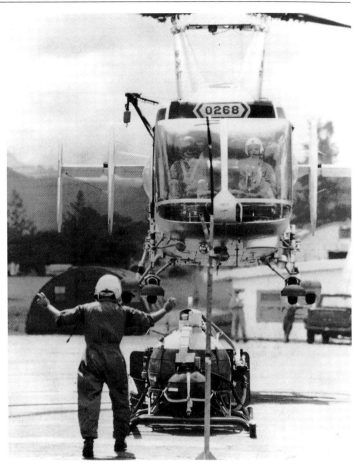

Guided by a ground crewman and a positioner pole, a Huskie hovers up to its FSK during a scramble mission. (U.S. Air Force)

Wright-Patterson AFB, Ohio. The unit was manufactured by the Surface Combustion Corporation, Columbus, Ohio.

The FSK (nicknamed "Sputnik" since it resembled the space satellite) weighed 1,000 pounds and consisted mainly of a globe, pressurized air tank and 150 feet of hose. The globe was a combination water and foam fluid tank. Its 78-gallon water capacity combined with 5 gallons of foam-producing concentrate to yield 690 gallons of foam, enough for 55 seconds of use. A firefighter opened a valve which released dry air or nitrogen to a pressure reducer, then into the water/foam fluid, forcing the mixture through a 1.5 inch fabric-covered, rubber-lined hose. When released through the nozzle into the atmosphere, it expanded to eight times its former volume and became foam.

At the onset, many types of extinguishing agents were tested, with mechanical foam far surpassing others when used in conjunction with the Huskie's rotor wash. The foam itself was a protein mix made from sheeps' blood, which was heavy, sticky and extremely foul smelling. In later years, Aqueous Film-Forming Foam (AFFF), called "light water," was developed as a firefighting agent, which, when heated, formed a protective air-tight film over a flammable liquid fire. Light water's low surface tension also allowed it to penetrate burning materials deeper.

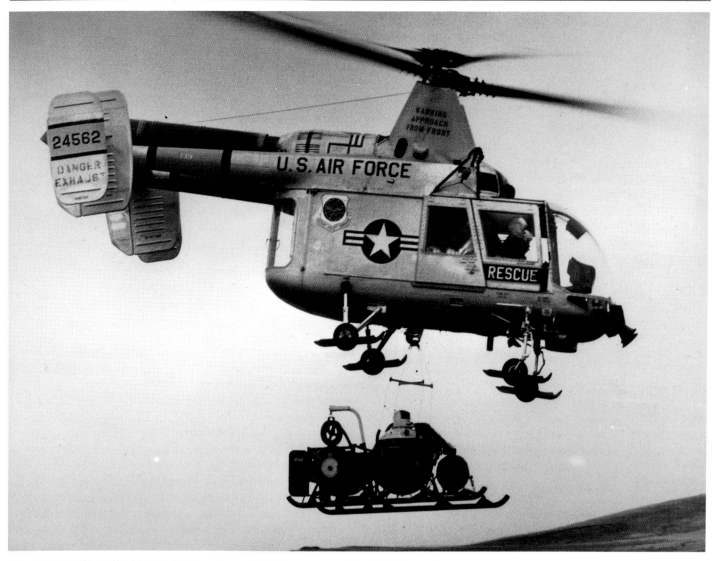

**This HH-43B lifts off with the final version of the fire suppression kit. (Kaman)**

During the mid 1960s, a modified version of the FSK was placed in service. The improved skid-mounted kit featured a hose reel with non-collapsible "hard" hose, which was stronger under extreme pressure and heat. Though 250 pounds heavier than its predecessor, the updated FSK featured vast improvements, which included components modified to improve reliability and two stabilizing vanes which prevented oscillation during flight.

When not in use, the FSK was stored on a USAF MB-1 modified trailer chassis, which incorporated a gasoline-burning "booster heater" to prevent freezing of the water/foam mixture during cold weather alert. To aid in crash entry and survivor removal, a portable folding ladder was added to the FSK, and a crash-entry kit and K-12 gas-powered rescue saw were strapped to the Huskie's cabin floor.

To facilitate hookup of the FSK, a "positioner pole" was devised, which trimmed valuable seconds from the Huskie's response time. Topped with two sections of rubber hose and an offset arm, the pole could be adjusted for a FSK pickup from the trailer or on the ground. The technique had the pilot hover and creep up to the pole until he lined up white bands on each arm with his horizontal canopy framework. By the time the aircraft actually touched the pole, a ground crewman had hooked up the FSK, knocked down the pole and signalled the pilot up, all in a few seconds.

The FAA was so impressed with Air Force Huskies in the LBR role that they began evaluating similar techniques during the 1960s for commercial aviation. Effective as it was, the Huskie's days were numbered. On 4 November 1970, the Air Force, impressed with the track record of Bell's Huey series, submitted a contract for 30 crash-rescue HH-1Hs to replace the venerable HH-43. Deliveries began during October 1971 and were completed during 1973. Both HH-1Hs and twin-engine HH-1Ns equipped with onboard firefighting equipment augmented the LBR Huskies until they were retired in 1975.

# Chapter 8
# War in Southeast Asia

During the final months of World War II, the need for a rescue capability on land prompted development of the helicopter for rescue. The ensuing Korean war not only nurtured vertical rescue, but ensured the helicopter's status as the primary air rescue vehicle in Southeast Asia. However, before the helicopter would be recognized as a highly effective rescue tool, the concept would endure a troubled evolutionary period, during which the HH-43 was a pioneer in disproving many negative conceptions.

Though American casualties were inevitable after the first U.S. Air Force unit was sent to Vietnam (during October 1961), the establishment of adequate search and rescue forces received little attention. Air Rescue Service management had not planned for wartime search and rescue support, and there were the usual political complications. Of particular concern was that the existence of a search and rescue structure would have suggested potential casualties exceeding those normally associated with flight training—which was supposedly the limit of U.S. involvement. Further inhibiting a sufficient rescue capability were aircraft ill-suited for search and rescue in Southeast Asia's geography. The weak rescue aircraft inventory was the result of budget restraints and the rescue doctrine of the late 1950s, which forced the Air Rescue Service to relinquish most of its helicopters.

As the primary vehicle for local base rescue, the HH-43 survived the cuts. However, it was less than ideal for search and rescue beyond air base peripheries. As the number of combat sorties rose sharply during the early 1960s, rescue was performed by U.S. Army, U.S. Marine and South Vietnamese helicopter crews, who were ill-equipped, unavailable at times and lacked rescue training. Those conditions prompted the Pacific Air Rescue Center to lobby for air rescue forces in Southeast Asia, which, as expected, fanned the embers of rivalry between the Army and Air Force.

Meanwhile, expansion of the U.S. advisory role brought about the 2nd Air Division's assignment to the U.S. Military Assistance Command, Vietnam, during 1962, to assume primary responsibility for search and rescue operations. After a search and rescue study was completed during September 1963, the commander of the Air Rescue Service requested six Sikorsky CH-3C helicopters. The CH-3C, which had been introduced during June 1963, featured a farther range, larger payload and faster speed than the HH-43. However, the Air Force had committed them to drone recovery and the space program, leaving no alternative but to modify existing HH-43Bs for combat search and rescue. Since the HH-43F Model, modified for combat survivability, could not be ready before October 1964, the decision was made to send six stock HH-

The first HH-43 to arrive at Nakhon Phanom RTAFB in preparation for combat search and rescue missions into Laos is offloaded during June 1964. (John Christianson)

This HH-43B lost its oil cooler over Thailand during 1964. Local villagers kept the crew company until the arrival of another Huskie from Nakhon Phanom. Most of the aircraft's markings were sprayed over to avert easy recognition of its origin. (John Christianson)

This HH-43B was destroyed during the first Viet Cong attack on an American air base, and the first Huskie lost in the war. Serial number 59-1571 was among nine U.S. aircraft destroyed by the heavy mortar attack, which occurred on 1 November 1964. Three other HH-43Bs assigned to the detachment (s/n 63-9712, 9714 and 9716) were damaged. (Marty Jester)

Two shovels were added to the Bien Hoa detachment's FSKs in 1964 after an A-1H Skyraider crashed and nosed over in deep mud. His canopy partially open, the pilot drowned before the Huskie crew could dig him out with bare hands and rifle butts. (Marty Jester)

A Huskie pilot poses with a vehicle reserved for HH-43 alert at DaNang AB during 1964. (John Christianson)

43Bs to Southeast Asia as an interim measure to keep pace with mounting combat sorties.

To no one's surprise, infighting between Army and Navy headquarters in the Pacific delayed arrival of the aircraft until April. The many issues that surrounded the Huskies' arrival were resolved during May when the Joint Chiefs of Staff assigned the responsibility for air rescue in Southeast Asia to the Air Force.

Three aircraft were obtained from Pacific Air Force assignments, while another three came from stateside units. Though they were the first USAF Huskies committed to the war effort, they were not the first in theater; during late June 1962, a pair of HH-43Bs, along with a small number of Sikorsky H-19s, had been turned over to the Royal Thai Air

Two HH-43B crews on "cockpit alert" for an F-100 and F-101 jet shot down over Laos on 18 November 1964. The CIA air arm "Air America" made both pickups. Crewmembers were outfitted with sidearms and flak jackets. (John Christianson)

The swing-out armor panel in the pilot's doorway, nose antenna, additional windshield bracing and armor plate on the nose and hoist areas identify number 63-9712 as a conversion to HH-43F. The Huskie is on Bien Hoa's alert pad during October 1964. (Marty Jester)

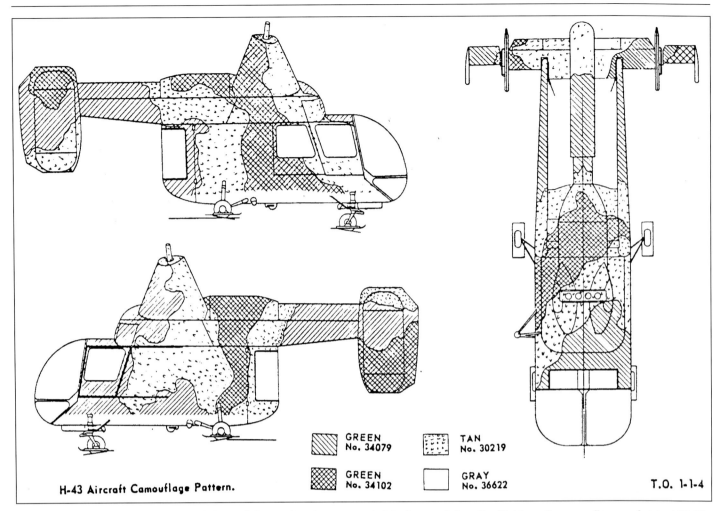

GREEN No. 34079    TAN No. 30219

GREEN No. 34102    GRAY No. 36622

H-43 Aircraft Camouflage Pattern.    T.O. 1-1-4

**U.S. Air Force Technical Order 1-1-4 used these drawings as a guide for applying the "tri-tone" camouflage paint to HH-43s based in Southeast Asia.**

Force for search and rescue. USAF pilots accompanied the aircraft as advisors to the RTAF.

As the war in Laos escalated during early 1964, U.S. reconnaissance flights resumed under the code name "Yankee

**A Huskie crew performs a rescue simulation at Bien Hoa AB during September 1964. An A-1H Skyraider hulk was used in the fire pit. (Gene Traczyk via Marty Jester)**

Team." Since the capture of an American airman in "neutral" Laos could cause an international incident, search and rescue became paramount. After the Air Force was granted permission from the Thai government to launch search and rescue missions into Laos from Thai bases, a pair of HH-43Bs from the 33rd ARS at Naha Air Station, Okinawa, was diverted to Nakhon Phanom Royal Thai Air Force Base (RTAFB). Nakhon Phanom, on the Thailand-Laos border, became better known to airmen as "Naked Fanny," or simply "NKP." The HH-43 contingent, which was originally destined for Bien Hoa Air Base (AB), South Vietnam, in compliance with the May directive, arrived at NKP on 20 June.

Despite a complete lack of working and living facilities, the HH-43 section was operational within four days at a clearing matted with a pierced-steel-planking (PSP) runway. Fuel arrived in a bladder via C-123 or C-130 cargo aircraft. Less than 200 personnel were assigned to the site, which initially was only a radar installation and rescue helicopter base. Two HH-43B crews pulled continuous alert for the entire five-month period the unit was at the site.

HH-43s in Southeast Asia flew with their rear clam-shell doors removed and a nylon cargo net stretched across the opening. This HH-43F of Det. 6, 38th ARRSq occupies the alert pad at Bien Hoa during 1971. (Author)

Few aircraft that ventured into the vicinity of Nakhon Phanom-based HH-43 crews were immune from this mischievously-applied "zap." (David Wendt)

Though the HH-43's presence was a morale booster for U.S. air crews, its limited range forced fliers to attempt to nurse their damaged aircraft to within 50 miles of NKP. Aircraft flown by Air America (the air arm of the Central Intelligence Agency) were available to fill the gap, but the intensifying air war over Laos quickly overwhelmed rescue capabilities. Air America helicopters took up the slack for the short-range Huskies throughout 1964 and 1965.

After the Gulf of Tonkin incident during early August 1964, the arrival of additional air units placed increased demands on the Air Rescue Service. In response to the urgency, HH-43Bs of Detachment 4 of the 36th ARS at Osan AB, Korea, were sent to Takhli RTAFB. As rescue forces in the Pacific became spread thin, local base rescue units in the U.S. were notified of impending deployments to Southeast Asia. Since the Pacific Air Rescue Center did not have the resources to keep pace with the rapidly heating war, five U.S.-based detachments were deployed to the war zone on a temporary basis. Detachment 10, of the Eastern Air Rescue Center at Maxwell AFB, Alabama, received orders on 6 August to leave

Designed by Walt Disney Productions for Det. 7, 38th ARRSq at DaNang AB, this insignia combines the Pedro call sign with the HH-43's dual role of fire suppression and aircrew recovery.

Considering the size of the HH-43, the three-color camouflage scheme specified for Air Force aircraft in Southeast Asia usually used only two colors (one shade of green and tan) on the helicopter. Serial number 60- 0278, seen here at Tuy Hoa AB during 1967, was destroyed two years later. (Carl Damonte)

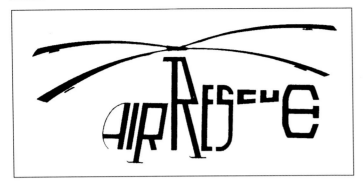

Captain Dennis Oleon of Det. 12 at Utapao RTAFB, Thailand, designed this distinctive insignia when his unit sought an emblem that best represented their unit and mission.

This HH-43B, serial number 59-1590, was painted an unusual camouflage of gray and black at Phan Rang AB during 1968. The aircraft was lost during 1969. (Hank Reed)

immediately. After working through the night dismantling and loading their two HH-43Bs aboard two Douglas C-124s, the unit arrived at Korat RTAFB, where they formed Provisional Detachment 4, which provided local base rescue for F-105s and the RTAF flight school.

A similar rushed response to a classified message was occurring at the opposite end of the country on 6 August. Detachment 20, of the Central Air Rescue Center at Minot AFB, North Dakota, made hasty preparations for the Pacific hop, but without their aircraft. Since the unit's HH-43s were close to scheduled replacement of major components, the detachment at Grand Forks AFB, North Dakota, was notified

to ready their Huskies. The personnel arrived at DaNang AB on 10 August, followed by the two HH-43s (S/N 62-279 and 280) aboard a C-124 on the 12th. A third Huskie (S/N 62-4510) arrived shortly thereafter from the detachment at Glasgow AFB, Montana. The contingent, which formed Provisional Detachment 2, became the first HH-43 unit in South Vietnam. The two HH-43Bs, which ended up at Phu Cat AB, served in Southeast Asia longer than any others; the third, number 62-4510, was shot down over North Vietnam.

Personnel attached to Provisional Detachment 1 during August 1964 were drawn mainly from Detachment 4, Western Air Rescue Center at Paine Field, Washington. During its initial assignment at Bien Hoa AB, the unit flew 142 combat support missions. Personnel from units based at Glasgow, Selfridge and McChord were also alerted to deploy.

On 9 September, the Air Force Chief of Staff approved a plan to send six HH-43Fs, when completed, directly from Kaman to Southeast Asia. In the meantime, HH-43Bs based at Bien Hoa and DaNang Air Bases were transferred to NKP and Takhli, allowing the temporary detachments to return to their U.S. squadrons. While the Huskie detachments at Korat and Takhli were restricted to local base rescue, the NKP-based unit was given the search and rescue mission for Yankee Team flights. On 18 November 1964, those Huskies, using call signs "Pansy 88 and 89," participated in the first large-scale search and rescue operation of the war,

The HH-43F number 63-9711 before camouflage paint during 1965. It was lost during May 1967. (Barry Ellis)

which was also the first ARS helicopter sortie into Laos. The debut had the Huskies crossing the Mekong River into Laos, where they were met by a pair of U.S. Navy A-1 Skyraiders, with a HU-16B Albatross flying command and control. The HH-43Bs investigated a F-100's probable crash site, which turned out to be smoke from another source.

The crash was subsequently located and Air America was sent to recover the deceased pilot. When an RF-101 went down three days later, the Albatross requested aircraft other than HH-43s for the pickup. Again, Air America was sent in since their choppers could fly tracks to avoid trouble areas, versus the direct routing taken by the Huskies, which simply couldn't fly as far.

Despite their limited range, Huskies penetrated deep into Laos and North Vietnam to rescue downed fliers. The daring forays, which sometimes brought Huskie crews within 40 miles of Hanoi, were accomplished using a number of methods. A distance advantage was gained by positioning a Huskie in east-central Laos at Paksane and another farther south at Pakse. To extend operating range, fuel stocks were prepositioned at forward sites in the Laotian panhandle. Known as "Lima Sites," the refueling points allowed Huskies to leapfrog to the rescue area. Another crude but effective technique utilized a 55-gallon fuel drum lashed in the cabin and tapped into the aircraft's fuel plumbing. When the drum was empty, it was rolled out the back end.

DET 10 38TH ARRSq

PEDRO

FASTEST RESCUE
IN THE DELTA
OPEN NIGHT OR DAY

WIFE SEND A DEAR JOHN?
BAD HANGOVER?
BLUE MONDAY?
YOUR AIRCRAFT QUIT?
GOT A TIGER BY THE BALLS?
CHARLIE PUT ONE ON YOUR ASS?
CALL BINH THUY FOR AIR RESCUE
COST: A ROUND OF CHEER FOR
OUR CREW

**The HH-43 detachment at Binh Thuy used this business card to advertise their prowess at rescue.**

As air operations in Southeast Asia gained momentum and emerged from beneath a veil of secrecy, an official search and rescue structure was organized. On 1 July 1965, Detachment 3 of the Pacific Air Rescue center became the 38th ARS, which was given primary responsibility for local base rescue and aircrew recovery. The 38th ARS, with the aid of inter-service coordination, supervised and controlled the activities of all rescue forces throughout Southeast Asia.

The first pickup by HH-43 of a flier downed over North Vietnam occurred on 17 May 1965. Ground fire brought down an F-105 Thunderchief over the Song Ba River northwest of Vinh. The pilot bailed out safely, only to find himself descending into the midst of a large enemy troop concentration. Luckily, he landed in a dense bamboo forest on a hillside thick with undergrowth that hampered enemy forces trying to reach him.

**The national insignia on this Thailand-based HH-43B was on a removable placard. When removed, the aircraft was void of all identifying markings for missions into sensitive regions. (U.S. Air Force)**

**Dressed in war paint, a Detachment 13 HH-43B prepares to lower the final version of the FSK at Phu Cat AB during 1968. (John Christianson)**

Seen from the HH-43 cockpit, this F-100 landed on fire at Phu Cat AB on 19 December 1967. The crew of "Pedro 22" scrambled, deployed the FSK and had the fire under control before crash trucks arrived. (John Christianson)

Number 60-278, which was lost during July 1969, prepares to pick up its pararescueman. (U.S. Air Force)

Enroute to the area were two HH-43s, which had to detour around flak positions and fly through heavy rain. As they neared the target area, the crews saw tracers arc into the sky and fighter aircraft busy trying to suppress the fire. Just as an HH-43 pilot spotted the downed pilot's parachute, an electronic homing signal and smoke flare from the "Thud" pilot pinpointed his location. As one HH-43 went down for the pickup, its crew answered ground fire with M-16 rifles. The second HH-43 descended and orbited the lead chopper while

its crew traded blows with enemy fire, which included a .50 cal. machine gun.

The pickup HH-43 descended into the site until its thrashing blades barely cleared trees and foliage brushed the aircraft's belly. Due to the towering bamboo and dense jungle, it was impossible to maneuver directly over the downed pilot. The HH-43 ran out its full 100 feet of hoist cable, which the pilot reached only after beating his way through thick undergrowth. During that time, the HH-43 remained exposed in a

A pilot's low approach at low airspeed in a downwind condition caused the demise of this HH-43 in Southeast Asia during 1969. (U.S. Air Force)

A Pedro pilot's typical combat mission gear, which included flight helmet, flak jacket, armor chest plate, M-16 rifle, .38 cal. revolver and survival knife. The large knife was a personal non-issue item. Among the survival vest's contents was a survival radio (preset at a rescue frequency), spare batteries, water and a medical kit. (John Christianson)

Barely visible against rough terrain, a Pedro from Detachment 13 at Phu Cat AB hoists its medic aboard who had been searching for survivors of a C-7 Caribou, which crashed during December 1967. All 26 aboard were killed when the transport impacted the mountainside in bad weather. One of the Caribou's wing tips is visible at lower left. (John Christianson)

A Sikorsky HH-3E "Jolly Green Giant" arrives at Phu Cat AB during spring 1968. The HH-3E eventually replaced the HH-43 in Southeast Asia. (John Christianson)

A Pedro at the end of a long hard day in Southeast Asia. (John Christianson)

The Pedro pad at Phu Cat AB during 1968. (U.S. Air Force)

A Pedro hoists communication gear to the top of Phu Cat's new control tower during 1967. (John Christianson)

Wearing fresh camouflage, serial number 60-280 occupies the pad at Phu Cat AB. (John Christianson)

A HH-43 alert crew at DaNang AB during 1964. Number 62- 4510 was shot down over North Vietnam during 1965. (John Christianson)

hover over the deadly zone. "Rescap" fighters kept the enemy at bay as the pilot was reeled in. Once aboard, the Thud pilot drew his revolver and joined his rescuers in firing at the enemy. The mission, which required a flight of more than 200 miles over hostile territory (made possible by extra fuel drums), earned both HH-43 crews the Silver Star.

By the end of 1965, 25 HH-43s shared the rescue role with a handful of HH-3 "Jolly Green Giants." However, aircrew recovery was still not commensurate with intensifying enemy defenses. Finally, by the end of 1966, after improvements in techniques and equipment and much trial and error, combat search and rescue came into its own as a viable part of air operations in Southeast Asia. By that time, a total of 19 HH-43Bs and 10 HH-43Fs were in South Vietnam and Thailand. That year also marked a name change for the HH-43, to "Pedro," after the radio call sign originally used at Laredo AFB, Texas, prior to the war. The label stuck, and before long,

caricatures, complete with sombreros, began appearing at Pedro bases. Due to the hazardous nature of combat rescue, Pedro crews became very close-knit and displayed a great deal of camaraderie. The year 1966 also marked the introduction of camouflage for HH-43s. Best known as "tritone," the tan and green over light gray scheme became the standard for USAF aircraft during the war. Although the technical order outlining the scheme called for two shades of green, frequently only one was applied. Prior to the use of camouflage, Huskies had their markings sprayed over, however, large national insignias remained. Eventually, the aircraft were painted overall light gray as an anti-corrosion measure. Small national insignias on removable placards provided the option of concealing the aircraft's nationality during operations in "sensitive" regions.

A Pedro detachment normally comprised two aircraft, however, larger and busier bases, such as Korat, Takhli, Bien

A Detachment 11 HH-43B rescues Air Force Captain Leo F. Dusard III after he ejected from his F-100 over the South China Sea in 1968. (U.S. Air Force)

Conditions were anything but favorable when this Detachment 6 HH-43F picked up a seriously-ill sailor from the heaving foredeck of the USS CHEMUNG in the South China Sea. Captain James A. Darden fought a 30- knot wind to maintain a hover during the pickup. (U.S. Air Force)

A HH-43 rescues a fighter pilot who had bailed out just north of Tiger Island in North Vietnam. Clearly visible are the pilot's parachute and raft, which the HH-43 flight mechanic sank with rifle fire. (David Wendt)

Number 60-281, with a "horse collar" sling on the rescue hoist, at Korat RTAFB during 1967. (Ray Ford)

Hoa, DaNang, Cam Ranh Bay and Tan Son Nhut, often had three or four assigned. For combat search and rescue, a Pedro crew usually consisted of pilot, copilot, flight mechanic and pararescueman (PJ), who stood alert duty for immediate scramble. Other four-man crew variations comprised two pilots and two fire-rescuemen, or one firefighter and one medic, or one firefighter and the crew chief.

Normal procedure had two aircraft directed to a mission area, with one flying "cover," while the other descended from 3,500 to 4,500 feet (above small arms range) for the pickup. The vulnerable HH-43s were often escorted by Army helicopter gunships or A-1 "Skyraiders," which could provide suppressive fire while the HH-43s attempted rescue. Flights to the north dictated a fixed-wing escort. If terrain precluded a landing, a rescue hoist cable was lowered. If the person on

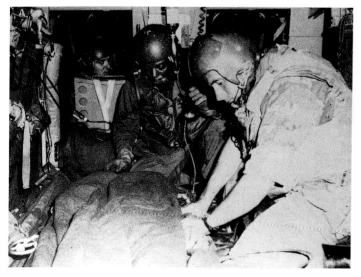

Concern shows on the faces of this HH-43F crew as the pararescueman treats a seriously wounded sailor rescued from a Navy patrol boat after it came under rocket attack. (U.S. Air Force)

A HH-43F lands at Bien Hoa AB during December 1964. The armored variants would later be camouflaged. (Marty Jester)

Lieutenants James Sovell and John Christianson pose with their HH-43B at DaNang AB during 1964. The Bear Paw skis were removed in an attempt to pick up a few knots of airspeed. The U.S. AIR FORCE lettering and all yellow markings were painted over. The sling or penetrator was attached to the rescue hoist enroute to the pickup point. (John Christianson)

the ground was unable to attach himself to the cable device, the PJ was lowered to assist.

In one instance, a PJ was lowered to help a downed pilot who pointed his revolver at the rescuemen during the descent. Obviously in shock, the pilot could not differentiate between the PJ and the Viet Cong, who were beating the bush nearby to find them. After several agonizing minutes, with the hovering Pedro serving as a sign post for the pair's position, the PJ was able to convince the pilot to get on the hoist. As the enemy closed in, the PJ had to wait for the cable to be lowered for his narrow escape.

HH-43s were known for speedy "feet wet" rescues of aircrew who declared emergencies over water. It was common for HH-43s to intercept a crippled aircraft, follow it to its ditching point and have the hoist device lowered to the aircraft— before crewmen were clear of the crash.

To effect rescue, it was often necessary for HH-43s to maintain an "extended hover" for minutes at a time, which, in combat, seemed like a lifetime. Often adding to the intensity of the moment was incessant and relentless enemy anti-aircraft or small arms fire. Pedro crews displayed incredible courage and perseverance in the face of odds that often favored a cunning and tenacious enemy. One favorite tactic had the enemy locating a downed flier, but instead of capturing him, they would stay at a distance and allow him to signal rescue forces. While aircraft were inbound, enemy troops set up offensive positions and "prepared to carve meat." Using the downed flier as bait, the ruthless enemy sprang the trap when the low and slow helicopter closed the distance for the pickup.

Prior to the arrival of the improved F Model, HH-43B crews devised makeshift protection for their "thin-skinned" Huskies. Improvised "armor" usually took the form of steel plates placed behind and under seat cushions. Ceramic inserts from body armor also found their way into various parts of the aircraft. A curved, half-inch thick, clear plexiglass deflector was positioned between the rudder pedals and windshield of one unit's aircraft.

When plans were drawn up to convert HH-43Bs into effective combat rescue helicopters (HH-43Fs), Air Force requirements specified survivability, endurance and a hoist unit with a cable long enough to reach beneath Southeast Asia's rain forests, which often towered well over 100 feet. The cable was lengthened from the standard 100 feet to 217 feet, 200 of which was usable. One severely wounded Army infantryman was especially grateful for the added length when a Pedro used all 200 feet to extract him from dense jungle. Attached to the cable was a "horse collar" rescue sling which became easily snagged in upper tree limbs. Kaman's solution, based on an early Navy design and introduced during 1965, was a "forest penetrator seat," which could be threaded down through the upper canopy of dense forests. The tapered device featured three spring-loaded arms which were lowered to form a seat. Successive improvements of the penetrator

The fire extinguisher is pulled clear and the FSK sling prepared as HH-43B number 60-282 scrambles at Takhli RTAFB during October 1966. (Dick Van Allen)

The tail fin tips of serial number 60-258 were painted red while it was assigned to Takhli RTAFB. (Dick Van Allen)

Number 60-278 provided rescue support for "Air Force One" (background) during the President's visit to Cam Ranh Bay during December 1966. (Dick Van Allen)

included a flotation collar, safety straps, and a shield to protect persons being raised through tree branches.

Survivability of the HH-43F was achieved mainly by the installation of 800 pounds of half-inch titanium armor to protect the crew and engine area. An armor panel added to the cabin interior's starboard side could be slid forward to cover the cabin doorway. Hinged armor panels, which swung out-

ward, significantly reduced both pilot door openings. Curved armor plating replaced the plexiglass on the lower cockpit windshield, and the windshield itself was reinforced with additional framing. Armor panels were also fitted to the cabin roof, cockpit floor and over the oil cooler and engine. Fuel tanks were made self-sealing, and a 150-gallon tank installed in the cabin increased the HH-43F's combat radius from 75

Between missions, HH-43s performed a variety of duties. Here, number 58-1845 moves a .50 cal. machine gun position at Cam Ranh Bay AB. This aircraft was lost with all aboard during 1968. (Dick Van Allen)

A HH-43B returns to Cam Ranh Bay following a mission during May 1967. (Dick Van Allen)

Crews of HH-43s became quite skilled at working with Navy patrol boats. This example makes a pickup during operations following a C-141 crash into the ocean off Cam Ranh Bay during April 1967. (Dick Van Allen)

Number 60-258 rests among cattle and expended ordnance at Kandi Bomb Range, Thailand during December 1968. (Dick Van Allen)

to 130 nautical miles. An increase in power was achieved by replacing the standard Lycoming T53-L-1 engine with the 1,100 shp T53-L-11A. Both VHF and FM radios were installed, and dual loudspeakers were mounted to the lower nose section. Although part of the improvement package included pintle-mounted M60 machine guns in the cabin doorways, it was soon discovered that they blocked the openings, thereby impeding rescue operations—crews reverted to sidearms for defensive armament.

Early in the war, HH-43 crews armed themselves with a variety of light weapons for protection and suppressive fire during missions. During 1964 a Huskie of Provisional Detachment 2 was notified that an Army UH-1 "Huey" helicopter was down just south of DaNang. Speed was vital since the last radio call reported the enemy very close and the crew almost out of ammunition. As the HH-43 neared the site, the Viet Cong revealed their positions by firing at, and hitting, the aircraft. The Huskie crew returned fire and called for assistance. Ground fire was too heavy to pull off a rescue, however, the HH-43 crew held the enemy at bay while an unarmed Army chopper dashed in to make the pickup. Thanks to the weapons aboard, the Huskie, which received nine hits, was credited with saving the downed Huey crew.

Since stateside crews were forbidden to bring new weapons into Thailand during the early period, they brought M-1 carbines and .38 cal. revolvers issued by their home units when they deployed. Before long, individual weapons of all types, including the more exotic Swedish-K and Thompson submachine guns, found their way into the niches of Huskie interiors. Some were scrounged by resourceful pararescuemen, while heavier pieces, such as BARs (Browning Automatic Rifles), came from CIA people in NKP City. The BAR was usually suspended from bungee cords in the aircraft's large rear opening, much like the arrangement used on Army and Marine helicopters. It wasn't unusual, on the other hand, for Air America crews on their way to rescues

deep in enemy territory, to stop at NKP to borrow BARs and carbines from HH-43 crews. At that time, crewmembers of CIA aircraft were not permitted to carry weapons.

After the first F Models became operational during early November 1964, the type was preferred for rescue where enemy fire was expected, and subsequently, a high degree of protection was desired. Studies of Huskie crashes in combat revealed one significant common factor—all the HH-43s were hovering within 200 feet of the ground, attempting pickups, when they were hit. In one incident, a pilot was killed, and as the copilot took over the controls, he encountered more intense gunfire. After landing at his base, 48 holes were counted in the Pedro, which was leaking fuel. Controls had been hit, however, the copilot reported that he had no difficulty flying the aircraft. Even before the war ended, statistics proved that the Huskie was more vulnerable to ground fire than other helicopters used in Vietnam. The sobering facts left no doubt that combat rescue was one of the most hazardous helicopter tasks.

Of major concern to Rescue headquarters was the Huskie's insufficient speed and range to reach airmen downed deep inside northern Laos and North Vietnam. Though the HH-43F stood a much better chance of surviving small arms fire, and both versions were superior high-altitude performers, hovering them in Southeast Asia's high density altitude was difficult, especially over mountains. The Huskie's limitations became a less critical factor as additional HH-3s, along with HH-53s (introduced during late 1967), were placed in service, although both continued to use Lima Sites for deep northern penetrations.

After the venerable Huskie paved the way for the bigger Sikorskys, it was used primarily for local base rescue with search and rescue as a secondary mission. A great deal of time was spent on support missions, as well as intercept missions, where Huskies scrambled aloft to meet aircraft that radioed for assistance (usually for "hung" ordnance), and escorted them to base. Huskies also became adept at rescuing wounded crewmen from Navy patrol boats in South Vietnam's Mekong Delta. Rescuing ground troops was another staple taken in stride by Pedro crews. After wounded were taken aboard the helicopter, they were treated by the pararescueman. At least one PJ was usually assigned to the flight crew of each ARS aircraft. As highly trained rescue specialists, they underwent rigorous and thorough training which enabled them to provide advanced medical treatment. When not braving enemy fire, Pedro crews frequently flew other missions that were equally hazardous. Hovering dangerously close to antennas and masts of ships at sea, Huskies delivered emergency supplies to stricken vessels and picked up injured sailors. The ships were usually in seas too violent for rescue by boat or Albatross seaplanes. Huskies were also called upon to transport patients from land to offshore hospital ships.

Besides combat missions, Southeast Asia-based Huskies carried on the Air Rescue tradition of humanitarian service wherever the need arose. Adjunct duties depended largely on a unit's area of operations and included perimeter patrols, beach patrols, supply flights to forward operating bases, support for explosive ordnance disposal teams, and the rescue of non-combatants, both military and civilian.

HH-43 crews of Detachment 7 at DaNang were on call to transport patients to the U.S. Navy hospital ship REPOSE (AH-16). Recalled from the Fleet Reserve during fall of 1965, the REPOSE had provided medical services to U.S. armed forces in Vietnam since February 1966. She was joined in Vietnamese waters by her sister ship, the USS SANCTUARY (AH-17), also equipped with a helicopter pad, during April 1967.

HH-43 crews also performed missions of mercy, usually on off days, as an extension of their traditional humanitarian role. Besides a busy average of nearly 100 alert launches every month, Detachment 3 at Ubon RTAFB flew mercy missions to villages in Thailand. As part of the 8th Tactical Fighter

Ready for anything at Phu Cat AB 1968. The pilot - Captain John Christianson and his aircraft - a HH-43B, both assigned to Detachment 13. (John Christianson)

Wing's civic action program, Pedros flew a medical team into remote areas on a weekly basis. An average of 100 people were seen on each trip.

A similar program was carried out in South Vietnam by Detachment 11 at Tuy Hoa AB. Each week, a HH-43 transported a medical team to the Cung Son Special Forces camp, which was a refuge for several communities of Vietnamese and Montagnards. There seemed to be no limit to the variety of tasks given the HH-43. During 1966, Rescue Headquarters canvassed HH-43 crews for volunteers for a three-month TDY (temporary duty) assignment. Prospective volunteers were told only that they would be operating at elevations of up to 19,000 feet. Returning crews were sworn to secrecy. The fact that they were linked logistically to Udorn RTAFB strongly suggests that they were involved with the placement of electronic listening posts atop mountains in southern China.

Huskies rescued four U.S. pilots on 2 March 1965 when six aircraft were shot down over the north. A fifth was plucked from the South China Sea by a HU-16B Albatross amphibian, while a HH-43 searched for the sixth near a North Vietnamese village. That HH-43's PJ descended the cable to look for the pilot, but found only his parachute, flight helmet and survival vest. A short time earlier, the pilot had been taken prisoner.

On another mission, while flying over the Gulf of Tonkin in search of a USAF fighter pilot who had bailed out, a DaNang-based HH-43F spotted the smoke signal of a Vietnamese Air Force pilot who had ditched his flaming A-1 Skyraider. The Pedro went down to within three feet of the wave crests, pulled the wounded pilot from the sea and started homeward. Enroute, it joined another HH-43F with a fighter escort that intercepted a distress call from a downed F-105 fighter-bomber pilot. Spotting a signal fire in the jungle, one HH-43 descended and hoisted the pilot 100 feet to safety.

Amazingly, one of four crewmen aboard this HH-43B survived when it was blown out of the air by a B-52 bomber explosion. The tragic incident occurred at Utapao RTAFB, Thailand on 19 July 1969.

In a mission flown by two Detachment 5 HH-43Fs, eight crew members of an Army CV-2B Caribou were rescued after their aircraft had a power failure nearly 60 miles from DaNang. The rescue of the crash-landed Caribou was complicated by a solid overcast over the crash area. Led to the scene by short radio counts from the Caribou, the HH-43 dropped straight down to pick up the stranded crew. After a safe landing through the cloud deck at DaNang, the Rescue Operation Center received reports that the Caribou crash site was under heavy mortar fire.

As a Battle raged near Phu Cat AB during early 1968, two HH-43B crews from Detachment 13 hovered in the gathering darkness to evacuate wounded from a jungle-covered mountainside. Tall trees and powerful wind gusts increased the hazard as one Pedro crew hoisted two soldiers aboard, then moved off to allow the other to pull up wounded. Army helicopter gunships kept the enemies' heads down during the evacuation, which was made even more difficult by poor visibility. Pedro crews dared not use their aircraft lights for fear of drawing enemy fire. As the HH-43s departed with the wounded, fighting intensified and artillery shells impacted the area, ensuring that the Pedros would be back.

When DaNang AB was attacked by rockets, mortars and enemy penetration teams during early 1968, a Detachment 7 Pedro displayed extreme versatility—the HH-43's trademark. When the rocket fire ceased, the Pedro was airborne to assess damage and search for casualties. From their vantage point, the crew observed impacting mortar shells and fighting along the perimeter. Flares from an exploding dump, along with blazing warehouses, aircraft and storage depots lit the sky over the huge facility. Under constant threat of exploding ordnance and enemy fire, the crew made numerous landings in search of casualties. During each landing, the firefighter radioed the exact location and type of each fire. When a large fire that was fed by a ruptured hose threatened a fuel storage area, the firefighter coordinated the fire suppression efforts of ground units with those of the helicopter. Meanwhile, the pilot took off and positioned the Pedro in a low hover just beyond the 40-foot high flames. He used the rotor downwash to blow the fire away from the threatened tanks, and to also clear a path for firefighters.

Soon afterward, the busy crew was notified that an aircraft was inbound and the base runways were covered with debris. No problem for the Pedro—it was hovered over one runway, allowing the downwash to clear obstructions. Not missing a beat, the Pedro was dispatched to the dispensary to evacuate a critically injured airman to the Naval hospital. As he was put aboard, the crew was advised that their flight path to the hospital was in enemy hands and heavy contact was occurring within 200 yards of the hospital. An AC-47 "Spooky" gunship flew escort for the flight, with the approach to the hospital covered by a pair of helicopter gunships. As if things weren't bad enough, the Pedro encountered another hazard—the entire area was blacked out, making it impossible to see the power lines that bordered the landing pad. As the gunships engaged the enemy, the pilot used landing lights to spot the wires and made a vertical descent before drawing fire.

The courage, professionalism and superior airmanship displayed by the Pedro crew during the three-hour episode saved the life of a critically injured soldier, prevented the loss of a multimillion dollar storage area, and kept the airfield op-

**Serial number 63-9714 after conversion to HH-43F at DaNang AB during July 1965. (Charles Burin via Terry Love)**

An outbound F-105 "Thunderchief" taxis past the HH-43 alert pad at Thailand's Takhli RTAFB during 1966. (Dick Van Allen)

erational for incoming aircraft. It was but one of countless actions by HH-43 crews, who exemplified the spirit of Air Rescue.

In one night during 1968, a HH-43 crew from Detachment 10 of the 38th ARRS flew six combat sorties in hostile territory within a three-hour period. Pedro crews from the same unit established a one-per-day average when they evacuated a wounded soldier from combat every day for eleven consecutive days. A crew from Detachment 1 started one day with a training flight, which was interrupted by a call to orbit for an inbound fighter carrying a bomb that failed to release. After the fighter landed safely, the crew spotted a gasoline fire that threatened a vital communications area. Using its rotor downwash, the Pedro controlled the blaze as firefighters moved in. The crew resumed their training mission, and again, practice was cut short when an Army helicopter made an emergency landing in an insecure area ten miles from the base. The Pedro picked up the Army pilots and flew them to base. That was followed by another intercept mission for hung ordnance. As they got in position, the crew heard a distress call from a Phantom jet crew ejecting near the base. They picked up the surviving pilot and conducted a search for the second one—all in a day's work.

The incredible achievements of HH-43s in Southeast Asia were not without loss; a number of crewmen were killed or wounded with at least 20 aircraft lost and many damaged. Operational duties as well as combat sorties exacted a heavy toll on HH-43 units. Maintenance sections performed admirably to keep aircraft flyable, often working long hours and under adverse conditions. During early 1968, maintenance personnel of Pleiku's Detachment 9 put a battle-damaged

HH-43 back in the air after it received more than 200 holes during a rocket attack on the base. The complete disassembly and repair included new rotor blades, tail section and nose glass, plus a new paint job.

U.S. Air Force Air Rescue strength in Southeast Asia peaked during the summer of 1969 with four squadrons operating under the 3rd ARRG. The 37th ARRS at DaNang AB and the 40th at Udorn RTAFB were responsible for search and rescue in Laos and North and South Vietnam. The largest number of HH-43s belonged to the 38th ARRS, which was headquartered at Tan Son Nhut AB and responsible for local base rescue, with aircrew recovery as a secondary mission. The 38th had detachments at 14 bases in South Vietnam and Thailand.

Though the war continued, the year 1970 marked the first withdrawal of rescue units from Vietnam. Throughout the year, the Air Force downsized and realigned Air Rescue units, a move which at first affected only local base rescue assignments. A major change took place on 1 July 1971 when the 38th ARRS was inactivated and its assets placed under the 3rd ARRG. When it became the first MAC unit to receive the Presidential Unit Citation during 1966, the 38th had already flown 8,700 combat missions. While rescuing 1,700 persons during its relatively brief life, squadron members were highly decorated, several were wounded and others were killed in action.

When it absorbed assets of the 39th ARRS, Detachment 4 of the 3rd ARRG temporarily became the largest local base rescue unit in the world until 8 July 1972, when it was redesignated the 56th ARRS. Air Rescue forces were further streamlined on 20 August 1972 with the 3rd ARRG, plus the

**Close examination of number 63-9714 at Bien Hoa AB during 1964 reveals a shot out windshield and bullet holes in the fuselage, copilot's window and engine exhaust. (Marty Jester)**

37th, 40th and 56th squadrons assigned to the 41st ARRW at Hickam AFB, Hawaii. Some HH-43s were assigned to local base rescue duty in the U.S., while others went to the Royal Thai Air Force to fulfill military assistance agreements.

The last HH-43s in South Vietnam were two transferred to the 40th ARRS from the 37th ARRS, when the latter was inactivated. The pair remained at DaNang AB to provide local base rescue during the Linebacker II bombing campaign. Pedro detachments remained in Thailand until January 1976, when the 3rd ARRG was inactivated and the last USAF Air Rescue elements were withdrawn.

Of a total of 2,039 combat saves accomplished during the peak period 1966 through 1970, by four rescue aircraft types, the HH-43 accounted for 888. By war's end, the Pedro was credited with more than 1,100 saves—not a bad record for a diminutive but tenacious helicopter initially purchased by the Air Force for firefighting.

## USAF HH-43 LOSSES IN SOUTHEAST ASIA

| DATE | SERIAL NO. | UNIT | KIA/MIA | CAUSE | INCIDENT LOCATION |
|------|-----------|------|---------|-------|-------------------|
| 11-1-64 | 59-1571 | PARC Det.1 | | ga | Bien Hoa AB |
| 6-2-65 | 63-9713 | | | gf | |
| 9-20-65 | 62-4510 | 39 ARRS | 1 KIA | gf | Vinh, North Vietnam |
| 7-8-66 | 59-1587 | | | | |
| 9-29-66 | | 38 ARRS | | | South Vietnam |
| 10-28-66 | 62-4511 | | 1 KIA | gf | Pleiku, South Vietnam |
| 11-12-66 | | 38 ARRS | 1 KIA | gf | Ia Drang, South Vietnam |
| 5-8-67 | 63-9715 | 38 ARRS | | unk | DaNang, South Vietnam |
| 5-21-67 | 63-9711 | | | | |
| 2-7-68 | 62-4525 | 3 ARRG | 1 KIA | gf | Kontum, South Vietnam |
| 9-27-68 | | 38 ARRS | 1 KIA | gf | Binh Dinh, South Vietnam |
| 10-10-68 | 58-1845 | 38 ARRS | 5 KIA | op | Ninh Thuan, South Vietnam |
| 1-26-69 | 63-9712 | | | | |
| 6-5-69 | | | | | |
| 6-28-69 | 59-1590 | | | | |
| 7-10-69 | 60-0278 | | | | |
| 7-19-69 | 59-1562 | 38 ARRS | 3 KIA | op | Utapao RTAFB, Thailand |
| 8-7-69 | 60-0282 | | | | |
| 12-17-69 | | | | | |
| 6-30-70 | | 38 ARRS | 2 KIA | | Laos |

Cause:  gf— ground fire
op— operational (non-combat)
ga— ground attack

NOTE: Ten are known to be combat losses from 1964 to 1969. Of those, eight were lost to anti-aircraft fire from 1965 to 1969 and two were destroyed during enemy attacks on air bases during 1964 and 1966. At least four were lost due to operational causes during 1969. These statistics represent the accurate data thus far available.

# USAF HH-43 ASSIGNMENTS IN SOUTHEAST ASIA

| UNIT | | LOCATION(S) | ACTIVATED | DEACTIVATED/RELOCATED |
|---|---|---|---|---|
| 33rd ARS Det. (Naha) | | Nakhon Phanom | June 1964 | Nov. 1964 |
| PARC (Prov.) | Det. 1 | Bien Hoa | Aug. 1964 | Nov. 1964 |
| | | Takhli | Nov. 1964 | Jan. 1965 |
| PARC (Prov.) | Det. 2 | DaNang | Aug. 1964 | Nov. 1964 |
| | | Nakhon Phanom | Nov. 1964 | Jan. 1965 |
| PARC (Prov.) | Det. 4 | Korat | Aug. 1964 | Jan. 1965 |
| PARC | Det. 5 | DaNang | 10-20-64 | |
| 37th ARRS | Det. 1 | DaNang | 1-8-66 | 11-30-72 |
| | | Nakhon Phanom | | |
| | Det. 2 | Udorn | 1-16-67 | 3-18-68 |
| 38th ARRS | Det. 1 | Nakhon Phanom | 1964 | 7-1-71 |
| | | Phan Rang | 1966 | |
| | Det. 2 | Takhli | 1964 | 11-15-70 |
| | Det. 3 | Ubon | 1-8-66 | 7-1-71 |
| | Det. 4 | Korat | 1-8-66 | 7-1-71 |
| | Det. 5 | Udorn | 1-8-66 | 7-1-71 |
| | Det. 6 | Bien Hoa | 1-8-66 | 7-1-71 |
| | Det. 7 | DaNang | 1-8-66 | 7-1-71 |
| | Det. 8 | Cam Ranh Bay | 1-8-66 | 9-15-70 |
| | Det. 9 | Pleiku | 1-8-66 | 7-1-71 |
| | | Nakhon Phanom | | |
| | Det. 10 | Binh Thuy | 1-8-66 | 12-20-70 |
| | Det. 11 | Tuy Hoa | 1-8-66 | 10-15-70 |
| | Det. 12 | Nha Trang | 1-8-66 | 7-1-71 |
| | | Utapao | | |
| | Det. 13 | Phu Cat | 1-8-66 | 7-1-71 |
| | Det. 14 | DaNang | 1-8-66 | 7-1-71 |
| | | Tan Son Nhut | | |
| | | Bien Hoa (rotational) | | |
| 40th ARRS | Det. 1 | Nakhon Phanom | 3-18-68 | 7-1-71 |
| | Det. 3 | Ubon | 8-20-72 | 8-20-74 |
| | Det. 5 | Udorn | 8-20-72 | 9-30-75 |
| | Det. 7 | DaNang | 11-30-72 | 2-10-73 |
| (Prov.) | Det. 8 | Takhli | 10-15-72 | 1-31-73 |
| | Det. 10 | Takhli | 1-31-71 | 7-30-74 |
| | Det. 12 | Utapao | 8-20-72 | 1-31-76 |
| | Det. 14 | Tan Son Nhut | 9-15-72 | 2-10-73 |
| 56th ARRS | | Korat | 7-8-72 | 10-15-75 |
| 3rd ARRG | Det. 1 | Phan Rang | 7-1-71 | 1-31-72 |
| | Det. 3 | Ubon | 7-1-71 | 8-20-72 |
| | Det. 4 | Korat | 7-1-71 | 7-8-72 |
| | Det. 5 | Udorn | 7-1-71 | 8-20-72 |
| | Det. 6 | Bien Hoa | 7-1-71 | |
| | Det. 7 | DaNang | 7-1-71 | 11-30-72 |
| | Det. 9 | Nakhon Phanom | 7-1-71 | |
| | Det. 12 | Utapao | 7-1-71 | 8-20-72 |
| | Det. 13 | Phu Cat | 7-1-71 | Nov. 1971 |
| | Det. 14 | Tan Son Nhut | 7-1-71 | 9-15-72 |

# Chapter 9
# Versatility Worldwide

Huskie missions were by no means limited to local base rescue, nor were they confined to the peripheries of U.S. air bases. Prudent officials were quick to capitalize on the Huskie's potential shortly after its introduction to the Air Force. A global dispersion of Huskie assets, though limited, resulted in astonishing achievements and countless lives saved. As more Huskies became available, detachments were formed, or relocated, to expand global coverage, and their missions became more diversified.

Huskies were there when a B-58 bomber crashed at a Paris air show. They were overhead directing firefighters when fire ravaged a district in Japan. They brought relief to thousands of flood victims in Mexico and Turkey. And they hovered over aircraft crash sites in forbidding mountain regions in preparation for the grim task of recovering bodies. They hovered protectively nearby as astronauts trained for moon exploration, and became high altitude performers during global mapping operations.

When the rain-swollen Chinwichon River washed away homes in South Korea during 1972, Det. 1 of the 33rd ARRSq at Osan AB sprang into action. The combined detachment, with three HH-3E and one HH-43F helicopters, saved 763 lives in 30 hours. That total far surpassed the total saves, including combat, credited to all services combined during the previous year. The detachment's HH-43F (s/n 60-252) set its own record by making 129 hoist pickups in less than six hours! The tension-filled Huskie sorties saw persons pulled from torrents, plucked from telephone poles, and in one case,

seven from a floating haystack, which disintegrated as the last survivor got on the hoist.

In a similar achievement, HH-43B crews from TUSLOG Det. 84 (Turkey-U.S. Logistics) at Incirlik AB, Turkey, saved the lives of 72 civilians and evacuated 344 more when the Tarsus-Mersin region was swept by floodwaters during December 1968. A few weeks later, the same detachment carried out another rescue and evacuation with one Huskie when a 15-foot snowfall isolated the 5,000-foot high mountain village of Arslankoy. The TUSLOG detachment earned the distinction of having saved 86 lives within a 30-day period, then an ARRS record.

Meanwhile, a large Huskie commitment in Southeast Asia necessitated the relocation of Huskie detachments elsewhere in the world. The 31st ARRSq at Clark AB in the Philippines set up helipads and fuel stocks at John Hay AB and Philippine Air Force radar sites to service HH-43s on missions to northern Luzon. Similar sites were established at NAS Cubi Point and NAS Sangley Point for far-ranging Huskie missions. Detachment 6, with three Huskies, was established at Kadena AB, Okinawa, which had previously been covered by the 33rd ARRSq at Naha. Detachment 6 stood alert for the Southeast Asia-bound 18th TFW and provided the only helicopter rescue presence in the Ryukyuan Islands.

Back in the U.S., the Huskie's involvement with the Military Assistance to Safety and Traffic (MAST) program was meeting with great success. As part of a government sponsored program employing military resources to aid civilian

An Ethiopian chief and clan pose with a HH-43B during the mapping mission in 1965. (John Christianson)

Comforts during the Ethiopian mapping mission were meager for this Huskie crew, which makes camp for the night. (John Christianson)

As the last load is pulled out of a deactivated site near Dira Dawa, Ethiopia during the U.S. mapping mission, the hookup crewman is hoisted aboard. (John Christianson)

communities, helicopters proved their worth, especially in sparsely populated areas, evacuating the seriously ill and traffic accident victims. Requests for MAST involvement were submitted to the federal government from more than 80 communities in 20 states. ARRS participation in the MAST program began during September 1970 when Det. 15 at Luke AFB, Arizona, and Det. 22 at Mountain Home AFB, Idaho, evaluated the project with the Department of Defense and Department of Transportation.

On the other side of the globe, a HH-43 of Det. 12 at Andersen AFB, Guam, was launched on a mission during

A HH-43B assigned to the Ethiopian mapping mission lifts a heavy load near Ginir. (John Christianson)

An injured worker, who fell inside a 150-foot tank tower and suffered a broken back, was hoisted by a Huskie through an 18-inch trap door in the tank's roof. (Kaman)

A Detachment 21 Huskie hovers protectively nearby NASA's LLTV conducting training for moon exploration at Ellington AFB on 29 January 1971. (NASA)

Just moments into its flight, the LLTV crashed and exploded, hurtling fragments which narrowly missed the HH-43. (NASA)

A H-43B based at Dover AFB, Delaware hoists the survivor of a light civilian airplane which crashed in rough terrain. (Kaman)

1971 after a Navy HU-16 Albatross requested a helicopter for assistance. The Albatross crew, in search of four fishing boats missing in high wind, spotted three wrecked boats and six survivors on coral reefs, near the base of a 400-foot cliff. With winds gusting to 40 knots, the Huskie pilot hovered into the wind at 25 feet to avoid waves, necessitating a backward hover to within 50 feet of the cliff. The pararescueman was lowered and the six survivors were hoisted to the top of the cliff—the Huskie then returned to the site and recovered two bodies.

As testament to the fact that Huskie missions had few climatic boundaries, those in the far north were accomplished under unique conditions. When a HH-43B of Det. 18 at Thule AB in Greenland completed 2,000 hours of accident/incident-free flying in the arctic, the milestone may not seem significant to the uninitiated. However, considering that chill factors dip to -70 degrees, batteries seldom stay fresh, whiteouts erase the definition between sky and snow, human flesh freezes in seconds, machinery bogs down and navigational aids were practically non-existent, the milestone is seen in a whole new light.

Not all of the Huskie's accomplishments were high drama. To its credit, a HH-43B was selected as one of the helicopters that participated in a competition held before Swiss Air Force officials to demonstrate load carrying capabilities at three different altitudes. At Emmen, Switzerland, the Huskie lifted off at 1,641 feet with a 4,480-pound payload. At the Mutthornhuette, a 3,000-pound payload was lifted from 9,250 feet, and at the Jungfraujoch, the HH-43B lifted off at 11,401 feet and hovered in ground effect with 16 men aboard.

Other notable high altitude operations included a goodwill mission where Huskies were called upon to aid in recovering the 41 victims of an aircraft crash in Peru. The wreckage was located at 14,700 feet on the slope of a steep mountain, 33 miles from Tacna. During the recovery operation, the HH-43 flew at altitudes up to 17,000 feet.

As part of preparations for America's Apollo flights to the moon, Detachment 21 was activated at Ellington AFB, Texas, during 1969 to provide rescue support of NASA's Lunar Landing Training Vehicle (LLTV) program. The HH-43 detachment also provided limited LBR coverage for other base flight operations.

Astronauts normally flew the LLTV below 500 feet at speeds of up to 35 knots, which enabled the vehicle's aerodynamics to best simulate lunar conditions. The Huskie's job was to fly close enough to the LLTV to rescue the pilot in the event of trouble, yet far enough away to allow him to maneuver. Having trained extensively with a mockup LLTV and the actual craft, the Huskie crew was keenly aware of the LLTV's fuel load—78 gallons of jet fuel and 800 pounds of hydrogen-peroxide—and its potential to create a violent explosion if mixed during a crash. A fire suppression kit could not be used since its foam would be ineffective on the unstable hydrogen-peroxide. The HH-43 crew, most of whom were seasoned returnees from the war in Southeast Asia, consisted of a pilot, two pararescuemen, a flight mechanic (in the copilot's seat) and a NASA flight surgeon.

On the morning of 29 January 1971, the crew of "Pedro 08" hovered in anticipation of a LLTV takeoff, which began routinely enough. However, a few minutes into the flight, the LLTV pilot experienced a flight control problem, causing the

Huskie crews flying MAST missions added to the long list of persons saved or aided by the ARRS. Here, a HH-43 of Det. 15, 42nd ARRSq, Luke AFB, Arizona carries out a MAST mission during 1972. (U.S. Air Force)

A H-43B brings a crewman ashore during a training exercise. (U.S. Air Force)

vehicle to tumble and accelerate backward out of control. The pilot ejected and the LLTV slammed onto the taxiway. To the Huskie crew's horror, it bounced in their direction and violently exploded in a huge fireball. With no time to get out of the way, the Huskie pilot held his hover through the shock wave as the LLTV's hydrogen-peroxide tank and other fragments streaked past the windshield. The HH-43 landed after the pilot/astronaut descended safely to the ground. Pararescuemen were at his side a few seconds later.

In stark contrast, some missions took the tough little helicopter well beyond the scope of rescue. Selected for their high-altitude performance, HH-43s participated in two mapping missions: in northern Australia and in Ethiopia.

Beginning in March 1962, the USAF Military Air Transport Service committed six aircraft, two of them H-43Bs, to measuring and mapping 88,000 square miles between northern Australia and Eniwetok Island. The remaining aircraft comprised three RB-50s and a C-54, and together the six aircraft formed Aerial Survey Team 7, which operated from Port Moresby, Papua New Guinea. Assisting the survey team were two small transport ships equipped with helicopter landing platforms. During a 27-day period, one Huskie flew 122 passengers and more than 40 tons of cargo while supporting six monitoring stations. Completed during 1965, the extensive

Huskies were used briefly to transport security policemen to and from remote guard posts at air bases. Unable to land due to obstacles, this H-43B uses its hoist to deliver a security officer to his post. (National Archives)

A total of 13 persons were rescued from a rooftop by this HH-43B during floods in Turkey during 1968. (Kaman)

project proved beneficial to world navigation, communication, transportation, power production, and even space exploration.

At the request of the Ethiopian government, the U.S. began mapping the East African nation of Ethiopia during late 1964. Working under a joint Ethiopian-American command, Detachment Provisional First (SAR) was formed with personnel and HH-43s drawn on a rotational basis from eleven Air Rescue detachments in Europe and Turkey. Personnel and a Huskie from the 55th ARSq at Kindley AB, Bermuda, also participated in the project, which was code named "King's Ransom."

Prior to the advent of mapping satellites, the procedure called for establishing HIRAN sites on mountaintops, which were interconnected by lines of position. When all the prominent peaks formed a network, USAF Photo Mapping Wing RC-130s took strip photographs based on earlier surveys.

Since the sites were manned for three months, the mapping wing was equipped with Sikorsky H-3 helicopters to supply them, however, they couldn't handle the altitude, which was usually from 10,000 to 13,000 feet. Therefore, the Huskie was brought in for a three-month test program, which ended up lasting two and a half years. The Huskies carried ten tons of equipment to each site when it was set up and brought in three-man teams to operate the station. Thereafter, they were supplied each month by Huskies that delivered nearly seven tons of supplies to sustain each station.

Flying at an average gross weight of 7,000 pounds, each HH-43B carried a 1,200-pound payload per sortie. Operating and living conditions were a far cry from those to which the LBR crews were accustomed. Sites varied from mesa-topped mountains offering "no sweat" landing conditions to mountain knolls barely large enough to accommodate the Huskie's landing gear. Pilots were confronted with strong winds, clouds and varying cargo weights. Huskie crews "jury-rigged" oxygen systems for flying above 10,000 feet, and fuel was often hand-pumped from drums, which had been lifted in earlier by Huskies.

During the summer of 1965, the Detachment Provisional First, true to its SAR tradition, broke from routine to join in a widespread search for two U.S. Army soldiers and an Ethiopian interpreter who were captured by bandits. At the end of a two-week search, the captives were released.

During both mapping operations, the Huskies were faced with every type of weather, terrain and altitude conditions imaginable, which provided challenges for man and machine. The fact that no HH-43s were ever out of commission during the long operations speaks well for the Huskies and their crews.

As a way of expressing their appreciation for the humanitarian and courageous rescue work carried out by Kaman helicopter crewmen, the Kaman Aerospace Corporation established a rescue award program in 1955. Providing the impetus for the awards was the severe New England flooding that year, when a Kaman test pilot and flight crewman

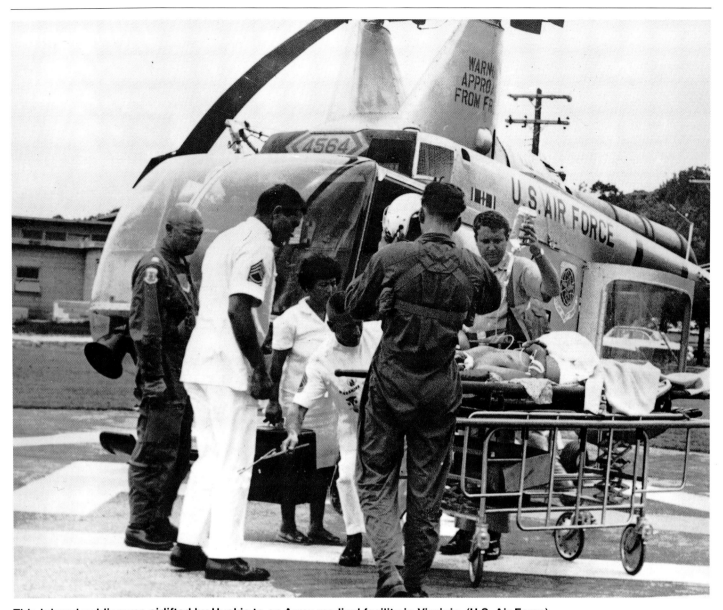

**This injured soldier was airlifted by Huskie to an Army medical facility in Virginia. (U.S. Air Force)**

rescued 13 persons with a HOK-1. Two types of awards were presented, the Kaman Scroll of Honor and the Mission Award, which recognized outstanding pilot and crew performance while conducting a rescue or mission of mercy with a Kaman helicopter. Kaman also presented awards to pilots when they logged 1,000, 2,000 or 3,000 hours in Kaman helicopters.

# Chapter 10
# Special Projects

It is no surprise that Kaman Aircraft Corporation's ingenuity resulted in a number of special helicopter projects. Recognizing the potential growth of the H-43B program, Kaman developed the K1125 helicopter as a private venture. With the improved variant, dubbed the "Huskie III," Kaman had hoped to meet growing national defense needs with a reliable, all-weather, medium-weight aircraft. It was also billed as an ideal helicopter for anti-geurilla warfare, missile site support and down-range recovery work on missile ranges.

The K1125 was basically an enlarged version of the H-43B, which combined all its dynamic components with a number of improvements. As the nation's first twin-turbine helicopter in the medium-weight range, the Huskie III was unveiled on 4 October 1962 at Kaman's Bloomfield plant.

The prototype was powered by two Boeing YT-60 engines, each of which developed 500 horsepower. Kaman listed its top speed at 138 mph, with a cruise speed of 121 mph. It had an empty weight of 5,230 pounds and a gross weight of 10,500 pounds. With 300 gallons of fuel, the Huskie III's range was 340 nautical miles, which could be extended to 715 miles with the addition of two 150-gallon fuel bladders. A single-boom empennage supplanted the twin boom arrangement, and a large ramp-type rear door, which incorporated stairs, replaced the clamshell doors. Other improvements over its predecessor included a larger cabin, seating 12 troops or 6 VIPs, cabin windows, fiberglass rotor blades, a 3,000-pound capacity cargo hook, and a chin pod housing electronic gear. Armament was available for the Huskie III in the form of machine guns and rocket pods.

Kaman proposed the K-1125, called the "Huskie III", as a replacement for the H-43. Despite its major improvements, the Air Force showed little interest in the design. (Kaman)

The Huskie III underwent evaluations for two years, but never gained acceptance by the Air Force. Undaunted, Kaman tried again; this time with a proposal in February 1969 for an improved LBR helicopter to supplement the HH-43 inventory. Kaman also advertised the type as an armed troop transport for counterinsurgency (COIN) operations. Called the K-700, the aircraft fared even less favorably than the Huskie III, barely proceeding beyond the drawing board.

Intended as the ultimate H-43 design, the K-700's power would be derived from twin T-400-CP-400 turbo engines driving the familiar intermeshing rotor blades. A stretched fuselage of nearly 42 feet would accommodate 14 passengers. Somewhat similar to the earlier proposed Huskie III, the K-700 was to have wheel landing gear, a "Vee" bullet-resistant windshield, self-sealing fuel tanks, provisions for external or internal auxiliary tanks, and an awesome array of armament, putting it on equal footing with contemporary gunships.

A vast assortment of features and options, too numerous to list, plus a guarantee by Kaman to deliver the first aircraft within 20 months, did not persuade the Air Force to purchase the K-700.

Artist's rendering of the K-700 helicopter which Kaman proposed for the Local Base Rescue and counterinsurgency mission. The artwork depicts the aircraft mounting an aft-firing maching gun and chin turret. (Kaman)

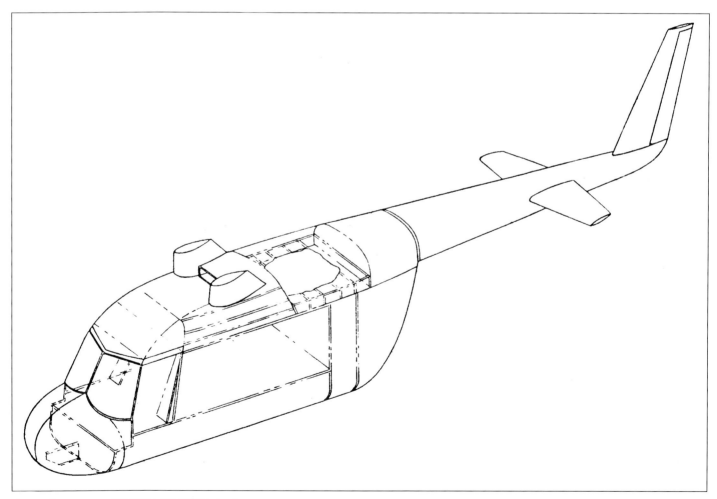

Kaman's proposed K-700 LBR/COIN design.

# Chapter 11
# Foreign Operators

At the beginning of the 1950s, when Kaman was hard pressed for production orders, the company tried to generate interest in foreign sales. A K-225 sent to Turkey, which was the first helicopter flown there, was purchased by the Turkish government. Foreign H-43 sales remained dormant until the end of the decade when a number of them were acquired by nations whose governments expressed a keen interest in the aircraft's capabilities, particularly its stability, versatility and high-altitude performance.

The first Huskies to be exported were six H-43Bs requested by the government of Columbia. As a nation whose history was marked by periods of internal strife, which severely damaged its economy, Columbia had to rely on American aid to shore up its aircraft inventory. The H-43Bs were sent under MAP provisions during 1961 to augment OA-10 Catalinas. Assigned to the Fuerza Aerea Columbiana (FAC—Columbian Air Force), they were registered FAC 251 through 256, which corresponded to their U.S. Air Force serials 60-251 through 256. The Huskies operated primarily from air bases at Melgar and Palanguero.

Duty with the FAC included transport of military personnel and equipment, search and rescue over the nation's coast, mountain ranges and Magdalena River, and helping to maintain order during civil unrest. The H-43Bs remained in Columbian service until 1968, when they were replaced by Bell Hueys and returned to the U.S.

Australia's Minister for Defense announced during late 1959 that light helicopters for search and rescue and liaison duties would be ordered. Kaman's H-43B was one of three competing designs for the Royal Australian Air Force/Navy helicopter. Though it lost to the Bell Huey in 1961, Kaman proposed local production of an interesting variant that featured float gear, external litter capsules, external fuel tanks and torpedoes.

The usefulness of helicopters in deploying and resupplying ground forces in counterinsurgency operations was realized at the onset of American involvement in Southeast asia. During 1963, the U.S. government supplied the Royal Thai Air Force (RTAF) with four HH-43Bs to supplement twelve Sikorsky H-19s. They were assigned USAF serial numbers

A Huskie of the Imperial Iranian Air Force undergoes tests in Tehran with a U.S. Air Force advisor aboard during August 1965. (John Christianson)

Huskies in Iranian service used a number of paint schemes, one of which was overall olive drab seen on this HH-43F at Shiraz in southern Iran. (MAP)

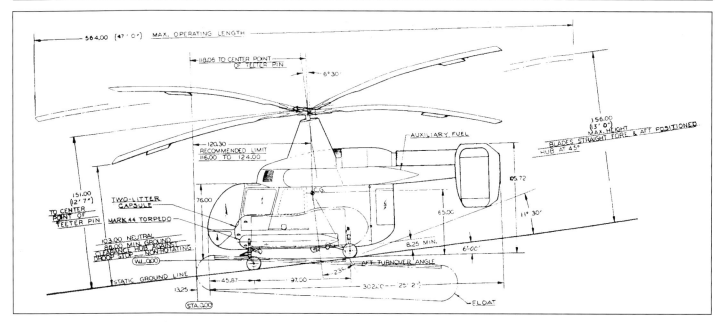

**Kaman alternate configuration proposal for Australian Huskie.**

60-290, 291, 292 and 61-2920, which became 6323 through 6326 in RTAF service. After initial assignment to the RTAF school at Don Muang Air Base, they were assigned to Squadron 31 of Headquarters Wing III at Korat RTAFB, where they shared duties with Bell UH-1s and Sikorsky H-34s. Designated H.5s (for the fifth helicopter type in RTAF service), the HH-43Bs served until the early 1970s.

The Pakistan Air Force took delivery of six HH-43Bs (s/n 62-4552 through 4557) for local base rescue. The Huskies wore "B" codes and camouflage schemes. Six HH-43Bs went to Morocco (s/n 62-5976 through 5979 and 62-12513 and 12514), which were later supplemented by four HH-43Fs (s/n 65-12755 through 12758) during the late 1960s.

The government of Iran was the recipient of the largest number of Huskies under foreign military sales. A total of 42 are believed to have been passed to that country, beginning with eight HH-43Fs during December 1964. Imperial Iranian Air Force (IIAF) crews received instruction alongside USAF HH-43 crews at U.S. bases, along with additional training in Iran by U.S. advisors. Duty with the IIAF consisted mainly of providing rescue coverage for aircraft operations at Mehrabod and Shahrokhi.

In December 1965, additional HH-43Fs were delivered, which helped form the Imperial Iranian Army Aviation Battalion. More arrived during June 1966, which were assigned to the Iranian Headquarters Supreme Command at Tehran for a variety of duties, which included assault troop maneuvers and transporting the Royal Family—the Shah of Iran became the only government head in the world to pilot a Huskie. Additional HH-43F assets were dispersed among bases at

The first H-43B turned over to the Thai government (s/n 60-290) wears RTAF and USAF markings shortly after delivery. (Robert F. Dorr)

A Huskie displayed at Don Muang RTAFB near Bangkok, Thailand. (MAP)

Columbian and American officials pose with one of six H-43Bs delivered under the MAP to the Columbian Air Force (FAC). (Oscar Forero R. via Marco Dini Bruno)

Columbian Air Force Huskie number 252 undergoes maintenance at the main helicopter base at Melgar. (Oscar Forero R. via Marco Dini Bruno)

Bushehr, Shiraz and Vahdeti, and missions became more diversified. Some Iranian Huskie crews worked in conjunction with the Gendarmerie to locate and capture outlaw tribal groups, while others flew mercy missions to isolated villages. During 1967, the Huskies began assisting Army ground forces in establishing mountaintop radio relay stations at approximately 11,000 feet.

The most noteworthy accomplishment during Iran's use of their HH-43Fs took place during 1968. A pilot of the Imperial Army Aviation Battalion at Isfahan rescued 18 troops at one time from a mountainside exposed to -20 degree temperatures and 45-knot winds. When it became obvious the

soldiers would freeze to death, all 18 were crammed into the Huskie, which made the flight without incident.

The final batch of Huskies delivered to Iran before relations with them ceased are believed to be eight 1963 vintage HH-43 conversions, most of which served in Southeast Asia. Throughout their service, Iran's HH-43s displayed a variety of color and marking schemes. Early arrivals and those assigned to Headquarters were originally painted overall white with black trim and wore the royal seal. Overall flat aluminum with orange trim was prescribed for some aircraft, while both IIAF and IIAA Huskies appeared in overall olive drab. None of the Iranian Huskies are believed to be flyable, much less in existence.

Huskies of the FAC were finished in flat aluminum and wore Columbia's national colors on the inboard tail fin. (Oscar Forero R. via Marco Dini Bruno)

# Chapter 12
# Survivors

Most H-43 helicopters, upon retirement from military service, were placed in storage at the Military Air craft Storage and Disposition Center (now called AMARC) in Arizona. The remainder were acquired by military and civilian museums. Most of those relegated to storage were purchased by Allied Aircraft Sales in Tucson, Arizona, which scrapped the majority of them. The few Huskies, HOK-1s, HTK-1s and HUK-1s that escaped the smelter found their way into the hands of operators who used them primarily for logging, firefighting and agricultural spraying.

Some commercial firms were fortunate to have obtained a substantial number of H-43s. Mosely Aviation Inc. owned a horde of Huskies it had acquired for spray work. At least seven of the firm's Huskies, along with five obtained during 1975, were used by the State of Washington's Department of Environment for firefighting and agricultural spraying. Likewise, Command Helicopters of Medford, Oregon, was able to latch on to a few HUK-1s for similar duties.

Thanks to the firm Timber Choppers Inc. in Bonners Ferry, Idaho, many of the sturdy rescuers were themselves rescued from the smelter and given a new life in the commercial market. Timber Choppers was a helicopter logging operation which got into the business of restoring Huskies during 1975. With an admiration for the type acquired through logging work, the company began hunting for and purchasing Huskies. By the time Timber Choppers contacted Allied Aircraft, all but 11 of the nearly 60 surplus airframes the firm had purchased were scrapped.

Since Kaman no longer supported the type, parts were difficult to obtain, until a lucky break in 1985. The Air Force had traded a cache of older aircraft parts to a company that restored military aircraft. Included among the parts were H-43 components, which Timber Choppers quickly purchased. The Air Force wanted four Huskies restored for displays, so Timber Choppers got the job and exchanged restoration work for Air Force stocks of Huskie parts as part of the deal. The Huskie specialists have completely rebuilt nearly 20 aircraft, along with nearly half that number for display purposes. During 1991, Timber Choppers refurbished two Huskies, which were sold to Kaman for training pilots in the K-Max helicopter.

As of this writing, nearly 30 flying H-43 helicopters thrive on the U.S. civil register, while others are known to be in operation around the globe. More than 20 serve as "gate guards" or reside in museums in the U.S. and abroad.

A handful of early H-43As survived into the 1980s as crop sprayers, such as this example which retained its rescue hoist. (MAP)

When Huskies were withdrawn from service at European bases during the mid 1970s, some were scrapped locally, while others were returned to the U.S. for storage. This pair of HH-43Bs awaiting shipment formed the LBR detachment at RAF Woodbridge. (MAP)

Before - serial number 59-1583 relegated to the boneyard. (Sid Nanson)

This rejuvenated Huskie, which began life as H-43B serial number 59-1576, is one of two owned by Kaman to train pilots in the K-Max helicopter. (Kevin Lederhos)

After - the same aircraft following rework by Timber Choppers. (Kevin Lederhos)

During the mid 1970s, this HH-43F was pulled out of storage for Detroit's Institute of Aeronautics. It was registered N62437 and later went to a private owner. (MAP)

One of a number of Huskies pressed into service by Washington's Department of Environment. Seen here during 1979, this example originally served the Air Force as serial number 59-1581. (via Terry Love)

Washington's Department of Environment eventually adopted this high-visibility color scheme for its Huskies. Former serial number 59-1544 appears here in 1986. (Douglas E. Slowiak via Terry Love)

Formerly serial number 59-1569, this Huskie was operated by Rogers Helicopters Inc. of Clovis, California. (MAP)

Command Helicopters was one of the few companies that acquired ex Navy HUK-1s for spraying, logging and firefighting. This example, minus its engine, appears to have seen better days. (MAP)

Having developed an appreciation for the Huskie's performance through logging work, and recognizing its growth potential, Timber Choppers Inc. went from logging to full time restoration of the type. This former HH-43F (s/n 62-4524) is seen at Fairchild AFB during May 1981. (Macpherson via Steve Miller)

This ex Navy HUK-1 (BuNo 146309) was rigged for spray booms for agricultural work in Oregon during 1972. (Dave Menard via Steve Miller)

Challenging as it may be to dress up the Huskie, there is no denying its performance as a stable platform for heavy work conditions. (John Wegg)

KAMAN H-43 HELICOPTER DISPLAYS

| TYPE | SER. NO. /BUR. NO. | LOCATION |
|------|------|------|
| HOK-1 | 129801 | New England Air Museum |
| OH-43D | 138101 | Naval Aviation Museum |
| OH-43D | 139974 | Pima Air and Space Museum |
| HOK-1 | 139990 | MCAS Tustin |
| HOK-1 | 139982 | MCAS Quantico |
| HTK-1 | 128654 | New England Air Museum |
| HTK-1 | 129313 | Naval Aviation Museum |
| H-43A | 58-1833 | Warner Robins AFB |
| H-43A | 58-1837 | New England Air Museum |
| H-43B | 58-1841 | Pate Museum |
| H-43B | 58-1853 | Warner Robins AFB |
| H-43B | 59-1578 | Kirtland AFB |
| H-43B | 60-263 | Wright Patterson AFB |
| H-43B | 60-289 | New England Air Museum |
| HH-43F | 62-4513 | Castle AFB |
| HH-43F | 62-4531 | Pima Air and Space Museum |
| HH-43F | 62-4532 | Travis AFB |
| HH-43F | 62-4547 | Buckeberg Museum, Germany |
| HH-43F | 62-4561 | Hill AFB |

NOTE: The New England Air Museum and Smithsonian Institution each are in possession of a Kaman K-225 helicopter.

This view of number N556D reveals the thorough refurbishment at the hands of Timber Choppers. (John Wegg)

This flashy Huskie, operated by Skyline Helicopters of College Place, Washington, once served the U.S. Air Force as a HH-43F. (MAP)

Former serial number 59-1592 flown by the state of Washington during the 1980s. (MAP)

This former HH-43 (s/n 60-284) was an Arizona resident during the mid 1980s. (MAP)

Following service with the Marine Corps and a commercial firm, this HOK-1 became a display at MCAS Tustin during the late 1980s. (Central Florida Aircraft Photographs)

The same aircraft after complete refurbishment wearing the "SU" tail code and warrior insignia which identified assignment to Marine Helicopter Training Squadron 301. The HOK-1 is displayed at MCAS Tustin, California, however, HMT-301 has since moved to MCAF Kaneohe, Hawaii, where it operates CH-53s. (Sid Nanson)

The same aircraft displayed in the markings of Marine Observation Squadron Six. The Bureau Number should read 139990. (Larry Ford)

Having seen better days, this OH-43D went on display at NAS Pensacola during the 1970s. (Steve Miller)

The Pima Air and Space Museum in Arizona boasts a dual Kaman display featuring a HH-43F and OH-43D. (Dale Mutza)

The Museum of Flight at Warner Robins AFB, Georgia is the only facility to display both "A" and "B" model Huskies, and is one of only two museums in possession of the rare H-43A. (MAP)

The older style horse collar sling is attached to the rescue hoist of this HH-43F at the Buckeburg Museum in Germany. (MAP)

Complete with helmets atop the pilot seats, this HH-43F is prominently displayed at the Castle AFB Museum. (Dale Mutza)

The H-43A and HH-43B display at Warner Robins AFB. (Kenneth D. Wilson)

The Hill Aerospace Museum at Hill AFB, Utah displays this well-maintained HH-43F. (Dick Van Allen)

This HH-43B rests proudly among a wide variety of rescue aircraft at the Air Rescue Museum at Kirtland AFB, New Mexico. (MAP)

# Chapter 13
# A Glance at the Future

During 1990 Kaman began a study of the commercial helicopter market and came up with an experimental aircraft that brought the company back to its roots after nearly half a century of building military helicopters. With its counter-rotating rotor technology, which originally set it apart from other manufacturers, plus a resurgence of interest in the stalwart H-43, Kaman decided to design a helicopter specifically for external load operations.

On 23 December 1991, the firm flew for the first time its "Multi-Mission Intermeshing Rotor Aircraft," better known as the "K-Max." The simple, rugged and reliable aircraft, which Kaman calls an "aerial truck," was designed specifically for

the logging industry. In view of the success with a handful of military surplus H-43s used for logging, Kaman was convinced that its unique intermeshing rotor/servo flap configuration was ideal for heavy lift operations—and the logging operators wanted another H-43. Striking a familiar chord with its past, Kaman proclaimed the K-Max's ability to double as an aerial fire truck, carrying a 700-gallon water bucket on the cargo hook.

Like its proven predecessor, the K-Max derives its lifting capability by diverting all engine power to hauling. Its single 1,800 shp Textron Lycoming T53-17A-1 gas turbine enables the K-Max to lift a 6,000-pound external load.

**This Kaman-owned Huskie was purchased from the New England Air Museum to train K-Max pilots. During its Air Force days, it carried serial number 60-289. (Kevin Lederhos)**

Finished in a striking scheme, this K-Max is seen in summer 1997 during operations at the "Hooper Fire" in the Angeles Forest of Los Angeles and Ventura County, California. The K-Max hoisted a 600-gallon water bucket. (Skip Robinson)

Kaman initially leased the single-seat aircraft to closely monitor their use before proceeding with sales. Deliveries began during September 1994. Pilots transitioning into the K-Max are first trained in H-43s, two of which were purchased from display collections and made flyable.

In a move that echoed its early corporate history, Kaman entered its K-Max in the U.S. Navy's Vertical Replenishment (VERTREP) demonstration competition during 1996. The K-Max won, and two aircraft began a two-month period in September, flying in support of Military Sealift Command's airborne cargo movement from ships in the Persian Gulf to demonstrate the feasibility of using commercial-leased helicopters for VERTREP operations.

Working in conjunction with Navy CH-46 helicopters, the pair of K-Maxs came off as natural, proving themselves 100 percent mission capable and available—in the finest tradition of the H-43.

With a Kaman test pilot at the controls, the K-Max undergoes Navy VERTREP trials off the USNS BIG HORN (T-AO 198) at anchorage off NAS Norfolk, Virginia during fall 1994. (U.S. Navy)

U.S. Navy pilots get their first look at Kaman's new K-Max helicopter in 1994. Charles Kaman (with hat) was on hand for the debut. (U.S. Navy)

# Kaman H-43 Series Production

| PURCHASING SERVICE | TYPE | SERIAL/BUNO. | NO. IN BATCH | REMARKS |
|---|---|---|---|---|
| U.S. Navy | XHTK-1 | 125446/125447 | 2 | K-225/Prototypes |
| U.S. Marine Corps | XHOK-1 | 125477/125478 | 2 | K-225 |
| | HOK-1 | 125528/125531 | 4 | |
| | HTK-1 | 128653/128660 | 8 | |
| | HTK-1 | 129300/129317 | 18 | |
| | HOK-1 | 129800/129840 | 41 | |
| | HTK-1 | 137833/137835 | 3 | 137835 became drone |
| | HOK-1 | 138098/138102 | 5 | |
| | HTK-1K | 138602 | 1 | |
| | HOK-1 | 139971/140001 | 31 | |
| | HUK-1 | 146304/146327 | 24 | |
| U.S. Coast Guard | XHTK-1G | 1239 | 1 | K-225 |
| U.S. Air Force | H-43A | 58-1823/1840 | 18 | |
| | H-43A | 58-5524 | 1 | became YH-43B |
| | H-43B | 58-1841/1860 | 20 | |
| | H-43B | 59-1540/1593 | 54 | |
| | H-43B | 60-0251/0292 | 42 | 6 to Columbia, 3 to Thailand |
| | H-43B | 61-2920/2922 | 3 | 1 to Thailand |
| | H-43B | 61-2943/2954 | 12 | |
| | H-43B | 62-4509/4565 | 57 | 6 to Pakistan |
| | HH-43B | 62-5976/5979 | 4 | all sold to Morocco |
| | HH-43B | 62-12513/12514 | 2 | both sold to Morocco |
| | HH-43B | 63-9710/9717 | 8 | most later sold to Iran |
| | HH-43F | 64-14213/14220 | 8 | all sold to Iran |
| | HH-43F | 64-15097/15103 | 7 | all sold to Iran |
| | HH-43F | 64-17557/17559 | 3 | |
| | HH-43F | 64-17682 | 1 | possibly to Iran |
| | HH-43F | 65-10647/10656 | 10 | all sold to Iran |
| | HH-43F | 65-12755/12758 | 4 | all sold to Morocco |
| | HH-43F | 65-12914/12915 | 2 | both sold to Iran |
| | HH-43F | 67-14769/14775 | 7 | first 6 to Iran |

# Appendix II
# Known Conversions to HH-43F

58- 1846
59- 1556, 1564, 1566
60- 0269
61- 2921, 2922
62- 4509, 4512, 4514, 4515, 4518, 4521,
    4524, 4526, 4527, 4528, 4529, 4530,
    4531, 4533, 4535, 4536, 4538, 4561,
    4562, 4563
63- 9712, 9713, 9714, 9715, 9716, 9717

# Appendix III
# USAF H-43/HH-43 Assignments

**44th ARRSq - EARC - Eglin AFB, FL (Robins AFB, GA)**
Det. 1 Loring AFB, ME (Det. 1 Thule AFB, Greenland)
Det. 5 Suffolk County
Det. 6 Andrews AFB, MD (Det. 59)
Det. 7 Seymour Johnson AFB, NC
Det. 8 Myrtle Beach AFB, SC
Det. 9 Shaw AFB, SC (Harmon AFB, Newfoundland)
Det. 10 Maxwell AFB, AL
Det. 11 Craig AFB, AL (Det. 53)
Det. 12 Moody AFB, GA
Det. 13 Brookley AFB, AL (Det. 58)
Det. 14 MacDill AFB, FL
Det. 15 Patrick AFB, FL
Det. 18 Thule AFB, Greenland
Det. 25 Eglin AFB, FL (48th ARSq, 44th ARRS)
Det. 42 Dow AFB, ME
Det. 43 Griffiss AFB, NY
Det. 44 Westover AFB, MA

**43rd ARRSq - CARC - Richards-Gebaur AFB, MO**
Det. 1 Glasgow AFB, MT
Det. 2 Minot AFB, ND
Det. 3 Grand Forks AFB, ND
Det. 4 Duluth Municipal Airport, MN
Det. 6 Kincheloe AFB, MI
Det. 8 Bergstrom AFB, TX (Selfridge AFB, MI)
Det. 9 England AFB, LA
Det. 10 Laredo AFB, TX
Det. 11 Laughlin AFB, TX
Det. 12 Randolph AFB, TX
Det. 14 Vance AFB, OK
Det. 15 Goodfellow AFB, TX
Det. 16 McConnell AFB, KS
Det. 17 Grissom AFB, IN
Det. 18 Little Rock, AR
Det. 21 Grand Forks AFB, ND (Ellington AFB, TX)
Det. 26 Selfridge AFB, MI

**42nd ARRSq - WARC - Hamilton AFB, CA**
Det. 2 Cannon AFB, NM
Det. 3 Kirtland AFB, NM
Det. 4 Paine Field, WA (Keesler AFB, MS)
Det. 5 McChord AFB, WA (Edwards AFB, CA)
Det. 6 Holloman AFB, NM
Det. 9 Portland International Airport, OR
Det. 12 George AFB, CA
Det. 13 Reese AFB, TX
Det. 14 Nellis AFB, NV

Det. 15 Luke AFB, AZ
Det. 16 Williams AFB, AZ
Det. 17 Davis-Monthan AFB, AZ
Det. 18 Webb AFB, TX
Det. 22 Mountain Home AFB, ID
Det. 24 Fairchild AFB, WA
HQ AFSWC Det. 1 Indian Springs AFAF, NV
3638th Flying Training Sq. Stead AFB, NV
40th ARRS Det. 4, 1550th ATTW Hill AFB, UT

## 40th ARRWg - AARC - Ramstein AB, Germany
Det. 1 RAF Alconbury AB, GB (Spangdahlem AB, Germany)
Det. 2 RAF Upper Heyford, GB (Leon AB, France)
Det. 3 RAF Lakenheath AB, GB (RAF Mildenhall, GB/Toul, France)
Det. 4 Ramstein AB, Germany
Det. 5 Hahn AB, Germany
Det. 6 Incirlik AB, Turkey (Cigli AB, Turkey)
Det. 7 Torrejon AB, Spain
Det. 8 Bitburg AB, Germany (Zaragoza AB, Spain)
Det. 9 RAF Wethersfield, GB (Moron AB, Spain)
Det. 10 Aviano AB, Italy
Det. 11 Incirlik AB, Turkey
Det. 12 RAF Woodbridge AB, GB
Det. 13 Spangdahlem AB, Germany
Det. 15 Zaragoza AB, Spain
Det. 20 RAF Woodbridge AB, GB

Beauvechain AB, Belgium
TUSLOG Det. 84 Incirlik AB, Turkey
TUSLOG Det. 153 Cigli AB, Turkey
Prov. Det. 1 U.S. Mapping Mission, Ethiopia
54th ARRSq Goose AB, Labrador/Det. 2 Harmon AFB, Newfoundland
55th ARRSq Kindley AB, Bermuda/Det. 1 Thule AB, Greenland
57th ARRSq Lajes AB, Azores
58th ARRSq Wheelus AB, Libya
67th ARRSq Moron AB, Spain

## 41st ARRWg - PARC - Hickam AFB, HI
Det. 3 Tan Son Nhut AB, South Vietnam
Det. 4 Taegu AB, Korea
Det. 5 Suwon AB, Korea
Det. 7 Misawa AB, Japan
Det. 11 Kunsan AB, Korea
Det. 12 Andersen AB, Guam
Prov. ARRSq 1646 Det. 1, Kunsan AB, Korea
31st ARRSq, Clark AB, Philippines
33rd ARRSq, Naha AB, Okinawa
36th ARRSq, Det. 1 Misawa AB, Japan
Det. 2 Yokota AB, Japan
Det. 4 Osan AB, Korea
Det. 8 Yokota AB, Japan
47th ARRSq, Fuchu AS, Japan Det. 9 Osan AB, Korea
Det. 6 Kadena AB, Okinawa
79th ARRSq, Andersen AB, Guam
NOTE: Data in parentheses refers to original designation or location.

*Also from the publisher*

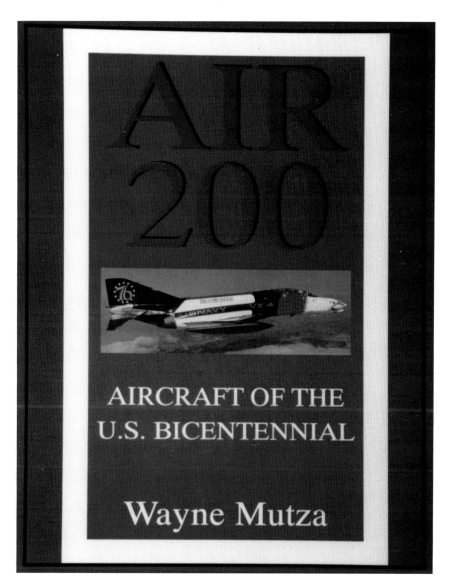

## AIR 200
## Aircraft of the U.S. Bicentennial
Wayne Mutza

Though Bicentennial aircraft schemes and markings, along with many of the aircraft of that period, have all but faded into obscurity, this colorful volume not only preserves their images, but reflects the spirit that prevailed during that historic period in U.S. aviation history. Presented here are the vivid, patriotic colors and schemes worn by aircraft during a time when the nation, still trying to purge itself of Vietnam, was swept with overdue patriotism; a time that compelled legions in civil and military air service to proudly "wave their flag" by decorating the aircraft that were their livelihood, or simply their hobby. A boon to aviation enthusiasts, historians and modelers alike.

Size: 8 1/2" x 11" over 190 color and b/w photographs
104 pages, softcover
ISBN: 0-7643-0388-0                                  $19.95

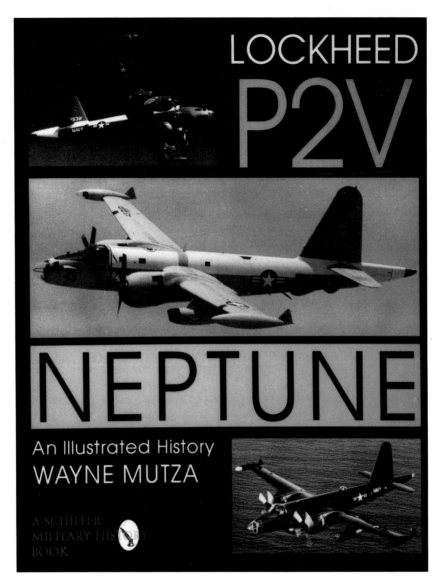

## LOCKHEED P2V NEPTUNE:
## AN ILLUSTRATED HISTORY
Wayne Mutza

This long overdue account provides an extraordinary amount of insight into the Neptune's lengthy history, beginning with its inception during World War II to the present day survivors. More than 1,000 examples were built, many of which thrive today as fire bombers and warbirds. Presented here for the first time are the many fascinating details describing Neptune service with non-U.S. air arms and obscure operations. Clearly evident is the in-depth research that makes this extensive volume accurate, detailed and readable.
Size: 8 1/2" x 11"
over 400 b/w & color photographs; 288 pages, hard cover
ISBN: 0-7643-0151-9                                              $49.95